ENVIRONMENTAL ETHICS IN THE MIDWEST

ENVIRONMENTAL ETHICS IN THE MIDWEST

Interdisciplinary Approaches

Edited by Ian A. Smith and Matt Ferkany

MICHIGAN STATE UNIVERSITY PRESS | *East Lansing*

Michigan State University Press
East Lansing, Michigan 48823-5245

Library of Congress Cataloging-in-Publication Data
Names: Smith, Ian A., 1975– editor. | Ferkany, Matt, editor.
Title: Environmental ethics in the Midwest : interdisciplinary approaches / Edited
by Ian A. Smith and Matt Ferkany.
Description: East Lansing : Michigan State University Press, [2022] | Includes
bibliographical references and index.
Identifiers: LCCN 2022001649 | ISBN 9781611864427 (paperback) | ISBN
9781609177126 | ISBN 9781628954807 | ISBN 9781628964745
Subjects: LCSH: Environmental ethics—Middle West. | Agriculture—Moral and
ethical aspects—Middle West. | Climatic changes—Middle West.
Classification: LCC GE42 .E676 2022 | DDC 179/.10977—dc23/eng20220415
LC record available at https://lccn.loc.gov/2022001649

Cover design by Amanda Frost
Cover art is AdobeStock. American Red Farm with
Chicago Skyline in Background by maksymowicz.

Visit Michigan State University Press at *www.msupress.org*

Contents

Preface

As in many other fields of "applied" ethics, environmental ethicists have come to learn that some of the best work begins down in the weeds. Valuable new insights for the field often emerge from close familiarity with the details of particular problems in particular places, not (only) from abstract theorizing conducted in isolation from the messiness of the world. At the same time, environmental ethics is a vast field with multiple dimensions. Environmental issues occur at multiple geographical and temporal scales, from local to global and from the present to the distant future, and commonly pose problems relevant to multiple fields of study, raising questions of biology, ecology, human psychology, culture, politics, and economics (and more). Where to train one's eye amid all this complexity?

For reasons of historical accident, the New England East and Mountain West figure large in the American environmental imagination. By the mid-nineteenth century, the New England transcendentalists (Emerson, Thoreau) pushed the concept and value of nature into the intellectual spotlight. A few decades later, battles waged by pioneer conservationists like Gifford Pinchot and John Muir in the West (over water rights, livestock grazing, damming of rivers, and flooding of valleys) led to numerous foundational environmental laws and institutions of American environmental governance and politics. One of the country's first societies devoted to environmental causes, Muir's Sierra Club, is named to honor the Sierra Mountains, and remains one of the most influential environmental organizations in the United States, if not the world.

Unsurprisingly, then, when Americans think about the environment and values associated with it, they often think about coasts or mountains. In fact, most environmental issues and values that get traction in the American national media, in anthologies on environmental ethics, in movies focused on environmental issues, etc., focus on issues in the West and East Coast regions. At the same time, collaborations between disciplinary ethicists and non-ethicists to address specific environmental problems in specific places remain rare, even while the interdisciplinary nature of environmental ethics is widely acknowledged.

Bucking this history, this volume will focus on environmental issues as they manifest in the American Midwest (though of course they can manifest in other regions of America and in other parts of the world). Made up of newly commissioned essays, the volume includes work by three collaborative, interdisciplinary teams, as well as scholars having diverse disciplinary and subdisciplinary backgrounds. The volume is thus entitled *Environmental Ethics in the Midwest: Interdisciplinary Approaches*.

The precise boundary and ethical culture of the Midwest is a matter of some disagreement. Although it is without ocean coasts and very arid deserts, the natural environment of the Midwest is diverse, including vast prairies, agricultural flatlands, dense urban centers, thick forests, Appalachian foothills, and even some sandy beaches along the enormous freshwater bodies of the Great Lakes. As the historical and present home of Indigenous nations, European colonists and immigrants, freed slaves, and immigrants from across South America and the globe, its social/cultural environment is equally diverse. Such diversity gives rise to social-environmental issues that, while not unique, transcend the historical preoccupation with wilderness conservation. The authors of this volume take on these matters (and more) in different ways, variously addressing the place and ethical culture of the Midwest: the ethical challenges of shrinking urban spaces; Indigenous women's leadership and struggles for control of waters and lands; antimicrobial drug use in hog farming; industrial watershed conservation and management; the impact of species conservation law on prairie species; and climate contrarian politics and science standards for climate education. The rest of this introduction provides a brief description of the chapters, plus a word on the use of the volume in teaching.

The agrarianism of the Midwest is often associated with an ethics of the stiff upper lip, what William O. Stephens calls "little s" stoicism. Tracing the place of nature in Roman "big S" Stoicism and relating it to Midwest agrarianism, Stephens describes a Midwestern Stoicism (big S) that he believes can help us cope with and ameliorate the environmental problems of today. Stephens also tackles the initially straightforward but, upon reflection, not so straightforward question of exactly what constitutes the geographical region of the Midwest. Moving to a discussion that is more focused on urban environmental issues, Benjamin J. Pauli and Levi Tenen attend to the challenges arising as Midwestern urban spaces contract under the strain of economic decline, population loss, and decaying infrastructure. Can urban planners deliberately shrink such spaces and so improve the environment and repair past environmental injustices without perpetrating new ones? The content of the next chapter—a discussion of Indigenous knowledge and values among Indigenous women and others in Standing Rock, North and South Dakota—provides a juxtaposition from the Western-centered (as opposed to Eastern philosophical) discussion of the first chapter. Matthew Meyer and Heather Ann Moody insert their discussion in the intersection of Indigenous feminism and women's politics of resistance as they consider questions like whether water is alive and what our proper relationship to water is.

Turning to issues more closely connected to human-animal relationships, J. M. Dieterle and Wade Tornquist's chapter provides a defense of the views that the use of antimicrobials in commercial animal agriculture should be heavily regulated, and that the use of prophylactic (non-therapeutic) antimicrobials should be prohibited. These views are not situated merely in environmental ethics, but also in bioethics, so the reader interested in bioethics in addition to environmental ethics will find this discussion beneficial. Also, the majority of hogs in the United States are farmed in the Midwest, so this is a pressing and relevant Midwestern issue. The next chapter considers issues that would especially interest the endangered species or biodiversity advocate. In his contribution, Ian A. Smith discusses the plight of the black-footed ferret and the monarch butterfly, both of which have central histories in the Midwest. Smith blends in a discussion of the philosophy of law with the philosophy of biology in his chapter, clearly demonstrating the interdisciplinarity evident in the volume.

Joel MacClellan's contribution follows suit and integrates environmental history with pragmatism in a discussion of dams along the Cuyahoga river in Ohio. Though this volume is not meant to be about case studies (in the words of one of our colleagues, Chris Diehm, such case-studies volumes can seem less relevant as time passes), MacClellan writes an informative and philosophically fruitful case study in such a way that it is not merely relegated to the past. For example, in encouraging his readers to move beyond thinking about environmental ethical dilemmas in terms of what has moral standing or intrinsic value, MacClellan introduces the reader to an alternative way of thinking about such dilemmas through the lens of a Deweyian pragmatism. Justin Donhauser uses a virtue-ethical centered framework to discuss another topic in Ohio, that of ecological restoration in the state, but not necessarily in the ways one might think. In pushing the dialogue of ecological restoration forward, Donhauser reveals cutting-edge research in the fields of environmental virtue ethics, assisted colonization studies, and novel ecosystem development in his discussion of restoration. Finally, Matt Ferkany discusses recent national science teaching standards for teaching climate change and how those standards haven't had as much traction in some states in the Midwest compared to some other states in other regions of the United States. Anthropogenic (human-caused) climate change is not just a pressing issue for the coasts of the United States, but for the Midwest, too. Ferkany shows this while addressing whether standards that would teach anthropogenic climate change indoctrinate students if they do not incorporate climate contrarian positions, a topic that engages the philosophical question of what constitutes indoctrination.

The essays of this book should be valuable to environmental ethics researchers, especially those looking for an example of how interdisciplinary environmental ethics can be engaged by teams of philosophers and other academics, and in an edited volume format. However, all are written to be accessible to intelligent nonexperts. Thus, this text should also be useful to upper-division undergraduate- and graduate-level teachers of environmental ethics, environmental studies, or related fields, especially instructors in Midwestern universities and colleges. In a course in environmental studies, for example, the text could serve as a standalone resource for engaging issues of both ethical theory and practice. In a full course in environmental ethics, the text could also be paired with an anthology or textbook in a subsidiary fashion, should an instructor want to add some Midwestern flair to the course, or the volume could be used as the focus for the course. One of the editors of this volume, Ian A. Smith, deployed the text in the second way in an upper-division undergraduate seminar on the volume. In using the volume as the focus of the course, the instructor in question could use a textbook to support analysis of the themes considered in the volume—for example, Joseph R. DesJardins's *Environmental Ethics: An Introduction to Environmental Philosophy*, 5th edition (2013)—as Smith did. One note about the format of the volume: at the end of each chapter a list of suggested readings is given so that students or professors interested in developing a research project based on the chapter in question have references to begin to work from; the lists contain central references found in the endnotes, but also references that are not directly cited.

Good environmental educators have long sought not only to teach their students about the environment, but to motivate them to care about it by teaching them in the particular place they live. However fuzzy the borders of the Midwest, we hope the focus on this region will help inspire researchers, general readers, and students who identify with it to care about and think reflectively on environmental issues and values in it. The Midwest is an amazing, complex natural and social place fully deserving of the attention of environmental thinkers across the disciplines.

Acknowledgments

The editors wish to thank Kim Hogeland, formerly at University of Kansas Press and now at Oregon State University Press, for generating the idea with Ian A. Smith back in 2017 to create this volume. We are forever grateful to her for helping us to plant the seed, as it were, for this project. The editors also wish to thank Julie Loehr, formerly of Michigan State University Press, in her role as acquisitions editor. We appreciate her helping us to land a great home for the volume at MSU Press. Special thanks are due to Catherine Cocks of MSU Press for her steadfast commitment as we brought the project home. Especially during the worst of the COVID-19 pandemic, Cocks provided essential help and encouragement even as our project's completion was inevitably delayed due to the pandemic. Finally, we thank the authors for the excellent work they agreed to contribute and share with the world here.

Midwest Stoicism, Agrarianism, and Environmental Virtue Ethics

William O. Stephens

Global environmental crises are acute and accelerating. Addressing these crises requires aggressive reforms to our systems, institutions, and ways of life. Effective remedies proceed from a holistic understanding of our environment, our place within it, and ourselves. The latter requires a good grasp of the character flaws that allow us to degrade the environment. We urgently need insightful environmental ethics to correct foolish, deeply engrained ways of life. In short, we need more than a little ecological wisdom to sharpen our perceptions and shrink our problems. My aim here is to elucidate the contours of this ecological wisdom in the tradition of virtue ethics.

The disposition of our ecological woes calls for caution. North America's environmental problems are so manifold and vast in scale that modesty requires narrowing the scope of concern here to one region. Where, then, to seek ecological wisdom? The region featured in this chapter is the Midwest. To begin there, however, immediately raises an epistemic quandary: Where *is* the Midwest? And how do we, or can we, *know* where it is? These questions are tackled in the beginning of this chapter. Next, I propose that the history of philosophy offers many sources of ecological wisdom. Specifically, I will argue that many aspects of the ancient Stoics' understanding of nature makes their philosophy a fertile field of ecological wisdom. The significance of nature in Stoicism occupies the next section. After that, I distinguish Stoic philosophers (Stoics) from stoical non-philosophers (stoics).[1] Nature's lessons for living a good Stoic life follow. A concern could be raised that such lessons in virtuous Stoic living may seem too theoretical to provide much practical guidance. To address this worry, I present several prominent ancient Romans who were lauded for their virtues and who worked the land. The virtues of these stoical Roman agrarians are discussed in the next section. Then I suggest that features of Roman Stoic agrarianism resonate with a contemporary instantiation of Stoicism in the Midwest. My thesis is that a Midwest Stoic agrarian, guided by an array of earthy virtues, will aim to engage in agricultural practices that *harmonize* with nature and support living in *agreement* with nature, while rejecting agricultural practices contrary to nature and propelled by vice. I conclude with reflections on the promise of Midwest Stoic agrarianism for fortifying environmental virtue ethics beyond this region.

Where Is the Midwest?

The seeming innocence of this question belies its deceptive vagueness and barely veils its incendiary potency.[2] Many who regard themselves as Midwesterners voice strong opinions on the matter. Heated disputes with a fellow philosopher born in rural Kansas and raised just over the border in rural Oklahoma have not shaken my certitude about my own answer. Born and raised a Hoosier, I learned in elementary school that my home state was the Crossroads of America. Crossroads are in the middle. The middle of the United States is the Middle West. Hence, Indiana is at or very near the center of the Midwest. Add to this syllogism the self-evident truth that Chicago is the greatest city of the Midwest. (Sorry, St. Louis.) Given Chicago's proximity to Indiana, support for Hoosiercentrism grows. More evidence comes from sociologists Robert and Helen Lynd, who found their statistical "Middletown" in Muncie, Indiana.[3]

If the question *Where is the Midwest?* is asking which U.S. states constitute the region, the U.S. Census Bureau answers: Indiana, Illinois, Iowa, Wisconsin, Michigan, Ohio, Missouri, Minnesota, Kansas, Nebraska, and the Dakotas. This tidy dozen might suit the intuition of a numerologist. My ordering of the first eight states describes a clockwise spiral. These eight states are so undeniably middle-western that since six of them have a coastline on a Great Lake, it seems apt to dub them the Great Eight.[4] The final four Plains States come last because they fringe states undeniably out west. Call these Plains States the Fringe Four. Moreover, consider the view of the research arm of the USDA, the Agricultural Research Service (ARS). The ARS divides the fifty states, the District of Columbia, and Puerto Rico into five regions. Of those five, the Plains Area groups the Dakotas, Nebraska, and Kansas with Montana, Wyoming, Colorado, and New Mexico. The ARS pushes Kentucky north to join the Great Eight.

Whether Kentucky is in the Midwest or the South is debatable. States are governmental entities that poorly cohere with cultural regions, much less bioregions. Settlement patterns and political history support the logic of including Kentucky in the Midwest since it allied itself with the Union during the Civil War. But Kentucky's heritage of slavery links it with the South.[5] Later, I will defend regarding the northern tier of Kentucky as marginally Midwestern.

Ambiguity about the bounds of the Midwest lingers, however, in another datum from the U.S. Census Bureau. Consider the concept of the mean center of the U.S. population. This is the point at which an imaginary weightless, rigid, and flat surface representation of the fifty states and the District of Columbia would balance if weights of identical size were placed on it, each weight representing the location of one American. This balance point is the mean center of population. The mean center of the U.S. population in 2020 was projected to be in Wright County, Missouri, within nine miles of Hartville.[6] Since the Midwest is known as the Heartland,[7] it is fitting that the town closest to the mean center of the U.S. population is named Hartville.

My defiant Okie chum insists that appealing to government bureaucrats to settle the issue of the location of the Midwest bespeaks naive statism. State borders are merely lines on a map, not ontological divisions. Census data are mere statistics. Map lines, road signs announcing, "You are now entering the State of So and So," and number crunching cannot settle the ontological issue of where the Midwest begins and ends. Motorists search in vain for road signs reading "Welcome to the Midwest." Locating the Midwest challenges not only geographers but also linguists, critics of music and literature, and social ecologists. It is a complex question about bioregions, the history of westward expansion, customs, shared norms of civility, and the prevalence of certain common courtesies. Such courtesies include holding doors for strangers, not cutting in line, and letting blameless motorists merge in front of you. Midwesterners share an accent and a dialect. Midwesterners favor certain foods and drinks. Clues about whether you're in the Midwest emerge when answering questions like: Is the tea sweet or not? Is the soft drink soda or pop? Is dinner cooked in a skillet or a frypan? Is the side dish grits, cornbread, or corn on the cob? Is dessert apple cobbler, apple crisp, or Apple Brown Betty? Bioregionalists may contend that the Midwest extends from the Ohio River west all the way to the Front Range of the Rockies, and from the Red River of the South north to Canada.

It is not that the seemingly innocent question of the location of the Midwest has no answer. It has several. Some convince more than others. The perimeter of the Midwest may have fuzzy edges, but Midwesterners are not relativists. So, having sketched key dimensions of this issue will allow me later on to adduce people, portraits, and phenomena of Midwestern provenance. A final verdict on Hoosiercentrism can wait. Before showing the relevance of Stoicism to the Midwest, I turn to explain the significance of nature in Stoicism.

What Is Stoicism and Why Was Nature Important to the Ancient Stoics?

Of all ancient Greek philosophies, I argue that the one with an especially intimate relationship with nature is Stoicism. The first premise of my argument is definitional. The earliest Greek Stoics defined happiness as "living in agreement with nature."[8] My second premise is that the Stoics conceived of the universe (*cosmos*) as a well-ordered whole. The English word *cosmetics*, the art of beautifully and meticulously arranging what is disordered, derives from the Greek word *cosmos* (world order). The Stoics believed that all events occur in harmony with providence—that is to say, they are willed by the divine. This theological understanding of nature is particularly robust.[9] Many scientists, environmentalists, and philosophers find it controversial, dubious, or outright misguided.[10] Some contemporary Stoics defend (their interpretations of) Stoic theology.[11] Nonetheless, the Stoics' view of nature as providential is tightly interwoven with their physics and ethics.[12]

The Stoics divided philosophy into three branches: physics, ethics, and logic.[13] They illustrated the relationship between these three branches with several similes. One simile is that if philosophy is like an orchard, then physics is the land and trees, ethics is the fruit, and logic is the surrounding fence. The fence does not directly concern us here.[14] On the other hand, my third premise emphasizes the significance of the analogy of the orchard's land and trees to Stoic physics and the orchard's fruit to Stoic ethics.

Let's inspect this fruit. Ancient Greek philosophers agreed that the goal of all human effort is an enduring state of well-being or flourishing activity they called *eudaimonia*. The Stoics believed that the purpose of philosophy is to achieve this goal by mastering the art of living. Stoic ethics educate us in this art of living in agreement with nature. Our biological nature involves in part using our sense organs to sustain our bodies. But we also naturally associate with each other. So, our social nature involves building interpersonal relationships, making friends, fostering a family, and creating a community. We recognize the affinity we have with all human beings and seek to live harmoniously with them. Thereby, we establish justice as the foundation of societal good. We also share many affinities with other animals and plants. Indeed, we share some affinities with insects, invertebrates, microbes, minerals, soils, lakes, streams, and seas. So, living in agreement with nature for Stoics includes living in congruence with landforms, watersheds, waterways, bioregions, and the entire biosphere. More on that later.

For a being with reason, living in agreement with nature means most especially living in agreement with reason, the perfection of which the Stoics call virtue. Virtue, they declare, is the only good because it alone is necessary and sufficient for self-fulfillment (*eudaimonia*). Conversely, the only thing that is bad and guarantees misery is the corruption of reason called *vice*. All else they consider neither good nor bad but *indifferent*. "Indifferents" are in themselves neither beneficial nor harmful because they can be used either well (virtuously) or badly (viciously). Most Stoics subdivided indifferents into the "preferred" and the "dispreferred." Preferred indifferents often but not always promote one's well-being. When they do, reason prompts Stoics to select them. Preferred indifferents include life, health, beauty, wealth, strength, and good reputation. The dispreferred indifferents are their opposites. It is often reasonable (virtuous) to avoid them, but occasionally reason dictates selecting them. A person's ethical character is measured not by having or lacking an indifferent, but by whether she makes good use of it or thrives without it. If she does, she is virtuous. Otherwise, she is vicious. Indifferents are like game equipment. A ball is neither good nor bad in itself. What matters is whether a ball player plays with (uses) a ball (one's back and arms, a plow, a plot of land, one's reputation, resources, etc.) well in the ballgame of life. The wise use of tools builds a happy, fulfilling life. The foolish, wicked use of resources makes a life miserable and unfulfilled.

So much for the fruit. How did the Stoics think about the land and trees, that is, physics? For the ancient Greek philosophers, physics (*ta phusika*) is the study of nature (*phusis*). According to Stoic physics, the cosmos is governed by an active principle called by many names: god, Zeus, Providence, Fate, Destiny, and Seminal Reason. This active principle (or agent) transforms matter into four elements: air (cold), water (wet), earth (dry), and fire/aether (hot). These elements combine to form objects. Stoics believe that god can be thought of as either the artificer or the orderliness of the cosmos. Stoics conceive of god not as anthropomorphic, but nonetheless as alive, immortal, rational, perfectly happy (self-fulfilled), and devoid of evil. This divinity provides for the cosmos and everything in it. Nature, Stoics explain, is either that which holds the cosmos together or that which causes terrestrial organisms to grow. Nature is a force moving by itself, producing and preserving in being its offspring in accord with seminal principles. Stoics describe nature as artisanal fire (*pur technikon*) equivalent to fiery or creative breath (*pneuma*). Because *pneuma* pervades the entire cosmos, all its parts are sympathetically interconnected. Stoics reason that this ubiquitous causal interlinkage is so seamless that all events are fated. Fate (*heimarmenē*) is thus an endless chain of causation whereby things exist.

Fate/Seminal Reason/Zeus/god shapes, crafts, and cultivates the universe and all its intricately, artfully interconnected parts. What do Stoics mean by "Seminal Reason"? Perhaps we can think of this active principle as a primitive precursor to our biochemical concepts of DNA and RNA. According to Stoics, Fate manifests itself biologically, we could say, as hot, breathy, creative "logic seeds" (*spermatikos logos*) causing all living things to bloom in characteristic ways homogeneously within their species. Like polynucleotides, we could imagine that these logic seeds stimulate growth, direct development, control homeostasis, regulate metabolism, and maturate all plants and animals.[15] Similarly, Stoic farmers labor to shape, craft, and cultivate their land in imitation of god/Zeus/Fate's crafting of the cosmos. Stoic farmers plow the rows, plant the seeds of their cereals and fruit trees, tend the fields, and, with the help of sunshine and rain—gifts of Providence from the sky—wait to harvest in good time. As Zeus provides sun, water, and earth for all residents of the cosmos, so Stoic farmers provide crops for their families and communities. Limited by their finite powers and mortality, Stoic farmers strive, as best they can, to emulate almighty, immortal Zeus. In working the land to produce sustenance for their fellows, Stoic farmers labor to be as providential as they can be in their locales. With nature's help they patiently transform tiny seeds into food.

Their farms are microcosms that both represent and are constituent parts of the living, deathless, well-ordered macrocosm. On a small scale, then, Stoic farmers exemplify Zeus/god/Seminal Reason, the Grand Cosmic Provider of all. Just as Zeus experiences a kind of bliss in tending the cosmos, so too do the farmers who, by farming, imitate the life Zeus enjoys.

Stoics do not conceive of *nature* (*phusis*) as we today usually think of the *universe*. From our contemporary perspective, we might be inclined to view nature as the universe seen through a telescope. Astronomers tell us that the universe is an inconceivably vast, unimaginably cold, infinitely expanding realm of distant galaxies, novae, nebulae, comets, black holes, quasars, dark matter, and dark energy. Such an infinite, black void punctuated by stars, superhot gases, planets with poisonous atmospheres, planets with no atmospheres, moons with volcanoes erupting ice, asteroids, and dust feels utterly sterile and terribly inhospitable. Nothing about this universe is welcoming. Such a boundless, frigid void is horribly hostile. Human beings with holistic understanding want not just a home (*oikos* = *ecos*) to live in, but to be part of the rational arrangement in nature. Stoics conceive of the cosmos not as inert, but as a well-ordered macro-organism of which they themselves are prominent parts. The living body of Nature with a capital N includes planets home to water in all three states, amino acids, habitats, sunshine, warmth, and life.

The portion of this living body of Nature most immediate to us is Earth's biosphere. Earth's biosphere is constituted of countless organisms of myriad sizes driven by biological processes synergistically embedded in the cycle of seasons, the diurnal cycle, the hydrological cycle, and evolution. These biological processes are also disturbed by air, water, and light pollution; soil degradation and erosion; acidification of the oceans; deforestation; desertification; accelerated extinctions of species; loss of biodiversity; and anthropogenic climate change (global weirding).

From the three premises above, I conclude that nature is of special significance in Stoicism. Consequently, it is entirely apt that many botanical examples take root in Stoics' arguments. Before we dig into those arguments, however, some concepts need to be clarified. Three groups of people must be distinguished: Stoic philosophers, stoical non-philosophers, and those merely resembling stoical non-philosophers.

Stoic Philosophers, Stoical Non-philosophers, and Stony Fools

First, there are those philosophers who defended by argument a specific, coherent system of thought. I described features of this system in the previous section. But this philosophical system, Stoicism, has exercised such a deep and lasting impact on Western history and culture that its legacy in English is the adjective *stoical*. To be *stoical* is to be impassive, to have calm, austere fortitude. One need not be a philosopher to have calm, austere fortitude. Consequently, here I will refer to philosophers dedicated to Stoicism as Stoics with a capital "S." I will refer to people with a calm, austere fortitude who don't philosophize as stoics with a small "s." The importance of differentiating stoics from Stoics is illustrated in the case of the (small "s") stoic farmer.

Consider a short medical journal article titled "The Case of the Stoic Farmer." The author, a Dr. Gracey, relates the story of an anonymous sixty-seven-year-old farmer from South Dakota. Call him Stan. Stan was visiting his daughter in Chicago when he was hospitalized because he was vomiting blood. Stan had a long history of pulmonary emphysema—a dangerous lung condition characterized by shortness of breath that may lead to impairment of heart action. For a long time, Stan had complained of a chronic cough and difficulty breathing. At the time he was hospitalized he was drowsy and audibly wheezing. Gracey writes: "It was hard to obtain a history. He was stated to have been brought to the hospital because his daughter noted that he was not eating and seemed to have a somewhat 'more unusual personality than normal.'"[16]

After many tests, intense questioning of Stan finally revealed that he had been vomiting frequently every day, a fact he had hidden from his daughter. For three weeks Stan experienced severe nausea and vomiting shortly after eating, but he would make a pretext of eating for his daughter's sake. After each meal he would go to the bathroom and secretly throw up what little he ate. Nowhere in the body of the article does Gracey label the patient a *stoic*. The good doctor merely emphasizes the importance of extracting a thorough medical history in order to avoid botching the diagnosis and muffing the prescribed course of treatment. "The lack of an important part of the patient's medical history was brought to light by the unusual initial blood-gas results and led to prodding the patient into revealing the source of the problem."[17]

Gracey considered his patient a "stoic" because he suffered in silence. Despite being unable to keep down a meal for three weeks, Stan resisted getting medical treatment. This "stoic" sufferer was nearly as reticent about disclosing his daily vomiting to a physician. Dr. Gracey regarded this secretly suffering man as a stoic with a small "s." But from the perspective of capital "S" Stoics, Stan was a fool. Stoics with a capital "S" strive for wisdom. Wisdom calls for emotional self-sufficiency but also honesty and candor. Wisdom does not abide hiding considerable physical suffering from loved ones. To calmly report one's own seriously unpleasant symptoms to a loved one is not to whine or whimper. Wisdom does not permit clamming up about one's ailments when being interviewed by health-care providers. If pride or embarrassment led Stan to pretend to eat normally while hiding his chronic vomiting, then he was being vain and silly. To be reticent about his retching was wretchedly foolish because it undermined the reasonable, natural norm of pursuing good health. Stan failed to reason well and act prudently, as capital "S" Stoics strive to do. Thus, to be silent as a stone, stubbornly hiding his illness, is not really stoical at all. Stan only superficially resembles a person of calm, fortitude, and prudence. Therefore, Stan was stubbornly stony, not stoical.

As an intently practical philosophy, Stoicism is a *practicable* way of thinking, desiring, responding to circumstances, and living available to anyone anywhere. Consider, for example, the following tale.[18] Long ago the horse of a Chinese farmer ran off. When his neighbors heard, they sought to comfort him, saying it was very unfortunate. But the farmer withheld judgment and said only, "Perhaps." The next day the horse returned along with seven wild horses. The neighbors congratulated the farmer for his good fortune to now own eight horses. The farmer replied, "Perhaps." The next day the farmer's son tried to tame one of the wild horses but was thrown from its back and broke his leg. The neighbors judged that this was very bad, but the farmer said only, "Perhaps." The next day state officials came to conscript young men into the army. Since his leg was broken, the farmer's son was rejected. The neighbors rejoiced, declaring this to be a great thing. Again, the farmer said, "Perhaps." The moral of the story? Nature unfolds as an inscrutably complex yet intricately integrated process. So, it is impossible to know with certainty whether any particular event, considered in isolation from the whole, is good or bad. You never know what the consequences will be of a putative "misfortune," nor the consequences of a putative "good fortune." While his epistemically reckless neighbors rushed to judge these events bad and those good, the Chinese farmer wisely restrained his judgments. His wisdom resonates with Zen enlightenment and Taoism.[19] For my argument, note that this is the story of a wise farmer. Wise farmers have learned to take what the Earth gives, when it gives, and however it gives, while letting go of what the Earth reclaims, when it does, and however it does. This is not thoughtless passivity. It is mindful adaptability. Many farmers lack it.

Nature's Lessons for Stoic Living

The most reliable and extensive sources of Stoic philosophy derive from the four Roman Stoics Seneca (4 BCE–65 CE),[20] Musonius Rufus (c.30–c.100 CE), Epictetus (c.55–c.135 CE), and Marcus Aurelius (121–180 CE). In this section I examine their reasons to think that nature instructs us how to live well.

Nature, in broad scope, is investigated by Seneca in his work *Natural Questions*. He studies in detail winds, clouds, rain, hail, snow, rivers, earthquakes, comets, meteors, rainbows, lightning, and thunder.[21] Probing study of the natural world is of serious intrinsic interest to Seneca, but he also emphasizes the great moral benefits of scientific understanding. He repeats the theme of wonder at the beauty of the world and assures his reader that study of the world of nature will yield knowledge of its divine ordering and of god himself.[22]

Seneca criticizes people who exploit the resources of Earth and seek to master it for their own pleasure and greed. The (Stoic) philosopher reveres the natural world and strives to understand it and the deity controlling it.[23] "For god did not make everything for human beings. How small a part of this vast creation is entrusted to us! He who manages all this, who created it, who laid the foundations for it all and surrounded himself with it, and who is the greater and better part of his creation, he eludes our sight and must be perceived by thought."[24] God entrusts the wise to contribute to God's grand creation precisely by acting rationally. Seneca's rationalism and naturalism foster his search for divine understanding. His love of nature merges with his dedication to education in the liberal arts. "The mind gains strength solely from liberal studies and from the contemplation of nature."[25] By "liberal" studies Seneca means those suited to a free person, not someone whose mind is enslaved by superstition. Free-thinking, scientific study of nature yields theological wisdom, which liberates our minds from the harmful passions of fear, greed, selfishness, vanity, and anger. These passions (*pathē* in Greek) are pathological diseases that obliterate sound thinking and good decision-making. These passions wreck our reason, ruin our minds, and rob us of happiness. Science, Seneca says, saves us. "When we have traversed the secrets of nature, when we have examined the divine, our mind must be set free from its ills and constantly strengthened."[26] Seneca doesn't separate science as a secular inquiry from theology as we today do.

For Seneca, the goal of studying nature is to liberate the mind. He accepts a popular belief in an earlier, simpler, ungroomed, uncorrupted age. He deplores his contemporaries' descent into vice from their virtuous forebears. Seneca contrasts the liberated mind that looks down from its cosmic vantage point on the narrowness of everyday existence at ground level with the dissolute deviants who look down to their base indulgences, self-absorbed with their depravities and carnal appetites. His investigations into nature aim to elevate us above these vicious habits. "Seneca impels his reader to look upward, to transcend ordinary life at ground level, to reach for cosmic consciousness."[27]

Naturally the farmer must look downward to the field that is to be tended. The farmer digs down into the dirt in order to raise crops upward to the sky. The ordinary life of Stoic farmers combines attention to the rain, sun, and storms above with working the soil beneath their feet. But in *Natural Questions*, Seneca contends that the mind of a Stoic can and should rise above all earthly worries, including (non-superstitious) worries about droughts, floods, erosion, freezes, and storms, all of which can damage crops. The mind and gaze of a Stoic can and ought to turn upward to achieve serenity and communion with the divine order of the heavens. In short, Stoic farmers cultivate their fields and minds simultaneously. Good agriculture nourishes good character and a good life. Prudent, level-headed farming is irrigated by worldly wisdom and fertilized by cosmic consciousness.

Seneca similarly links divinity with earthiness in his *Letters on Ethics*. He writes: "God actually comes *into* human beings. For excellence of mind is never devoid of God. Seeds of divinity are scattered in human bodies: if a good gardener takes them in hand, the seedlings resemble their source and grow up equal to the parent plant. But poor cultivation, like sterile or boggy soil, kills the plants and produces only a crop of weeds."[28] Again Seneca likens good gardening to divine cultivation. Improvement of the mind is a divine achievement. When virtues grow in a person's mind it resembles expert gardening that raises seeds into seedlings into thriving, mature plants. God orchestrates the cosmos so that, as we mature from children into adults, godlike reason blooms in us, enabling us to think and live well.

In an earlier section we saw how the fenced orchard simile illustrates the three components of Stoic philosophy. Seneca too gleans understanding of philosophical theory from the structure of trees. A subtlety in ethical theory—the distinction between the fundamental elements or basic principles (*decreta*) and the precepts or practical rules (*praecepta*)—he explains thus: "The difference between philosophy's principles and its precepts is the same as that between the basic elements of something and its branches. The latter depend on the former, which are both their causes and the causes of everything." Later in the same letter, botany again clarifies the dependency of precepts. "Leaves cannot flourish by themselves; they need to be fixed in a branch from which to draw their sap. In the same way, precepts on their own wither; they need to be fixed in a philosophical system."[29] Without their branches, leaves can neither grow nor thrive. Similarly, practical rules cannot guide action unless they sprout from a strong understanding of fundamental principles. Seneca uses arboreal morphology to impart a lesson in ethics.

Seneca draws five ethical lessons from contemplating nature

1. Beholding the beauty of nature teaches the orderliness of the world, an orderliness that is divine. Thus, study of nature yields theological wisdom.
2. Theological wisdom frees us from pernicious passions like fear, anger, and greed.
3. Admiring the marvels of the starry heavens above pulls our attention away from vain self-indulgence and vices of the flesh.
4. God implants in us the seeds of excellence, which we can cultivate into full-grown virtue, just as proficient gardeners nurture seedlings into thriving adult plants.
5. The structure of trees illustrates the relationship between basic theoretical principles and practical precepts.

Whereas for Seneca gardening serves as a useful analogy for ethical improvement, Musonius Rufus praises farming itself. Farming is an occupation no worse than philosophy, he argues, and it is perhaps even better for a man strong in body. He recommends earning a living from the land, if you own some, and even if you don't. Many people, he remarks, farm someone else's land to support their families.[30]

Because they work with their own hands and are industrious, some earn a very good living this way. The earth repays most beautifully and justly those who care for her, giving back many times what she receives. For someone willing to work, she supplies an abundance of all the things necessary for life and does so in a seemly and shame-free manner.[31]

Musonius declares that sowing, plowing, working vines, harvesting, and threshing are all tasks compatible with freedom and suitable for good people. Hesiod was not ashamed of being a shepherd but was loved by both the gods and muses. So, Musonius infers, only the decadent or soft would dare say that agriculture is unsuitable or shameful for good folk.[32]

Musonius implies that shepherding and certain other agricultural chores allow for leisure. He explicitly identifies the main benefit of farming as the abundant leisure for deep thinking and reflecting on the nature of education. This notion of the mind being freed to contemplate higher subjects than what lies at one's feet harmonizes with the goal of cosmic consciousness in Seneca's *Natural Questions*. Musonius seems to believe that many farming chores, as well as shepherding, require little exertion, and so these light tasks consume little time and thereby allow the mind to contemplate the better things and become wiser, the goal of every philosopher.[33]

Agrarian life offers many advantages. Musonius judges it better to be nourished from the earth than from some other source. (He also advocates a lacto-ovo vegetarian diet,[34] so presumably he rules out food from the sea, livestock, and wild game.) He judges it better to live in the country than the city (as sophists do), where false beliefs about what is good abound[35]; to be healthy by living outside than to seclude oneself in the shade; to be free and seemly by providing necessities for oneself than to get them from others. Therefore, Musonius concludes, earning a living from farming is noble, blessed, and favored by the gods.[36]

To the objection that time spent farming prevents one from philosophizing, Musonius replies that students of philosophy benefit less from attending a philosopher's formal lectures in an urban setting than by watching their teacher out in the country work and endure pain rather than ask for a handout. He believes that the philosophy student who works *with* his teacher down on the farm can simultaneously listen to him speak more convincingly about self-control, justice, and courage. Those who want to philosophize properly, he contends, don't need a lot of words or sophists bloviating and bandying about a big batch of theories. "Those who do farm work can learn the most essential and useful things, especially if they will not be working all the time but can take some breaks."[37] How authentically one walks (hoes, plants, prunes, etc.) one's talk matters. Musonius defends rugged training and philosophizing in *practice* performed by hardy, outdoorsy philosophers whose activities and habits display the virtues they propound. He steers youths away from soft, slick, decadent, citified, theory-mongering philosophers who don't do a lick of manual labor.

According to Musonius, progress in philosophy is measured by observed behavior. He affirms that all true lovers of philosophy would benefit from and relish spending as much time as possible in the country eating, drinking, sitting, and sleeping unconcealed, under the watchful eye of a good man.[38] Agrarian mentoring is just the ticket. Therefore, Musonius holds that farming fosters the virtues of industriousness, austerity, hardiness, courage, seemliness, freedom, self-reliance, nobility, self-control, and justice. So, whereas for Seneca contemplating nature is a fine cerebral activity productive of virtue, Musonius extols the physicality of working fields and robust, rustic ways of life as excellent means of becoming good.

Musonius's student Epictetus conceives of nature as theologically imbued. The divine Providence he sees at work in nature probably strikes contemporary sensibilities as unscientific and implausible. Epictetus believes that Zeus/god cares for human beings in many ways. These divine gifts include the rain that sustains trees bearing fruit. He states that when Zeus wants to be Rain-Bringer, or Fruit-Giver, or Father of men and gods, he only attains his goals and earns these epithets by benefiting the common good.[39] Zeus's gifts include not only life itself, but what sustains life, such as the vine and wheat, dried fruits, wine, and olive oil.[40] The will of Zeus/god is directed only at the global good of the entire world order, which includes the collective welfare of the cosmic community populated by all rational beings, mortal and immortal. For Stoics precipitation is not a phenomenon devoid of intention. Rather, it is beneficial on purpose.

Epictetus observes divine volition pervading biological processes. Moreover, he contends that the divinity watches a person's every action.[41] He argues for this by appealing to a doctrine of Stoic physics. The idea is that the physical cohesion of the parts of the Whole (the universe) is so seamless that what happens on earth is felt in sympathetic connection (*sumpatheia*) in heaven. The regularity with which plants flower, send out shoots, produce fruit, drop their fruit and leaves, and go dormant Epictetus attributes to god's commands.[42] He believes that the waxing and waning of the moon and the seasonal approach and recession of the sun correspond with terrestrial alterations from one opposite (for example, winter's cold, spring's wetness) to the other (for example, summer's heat, autumn's dryness). He thinks that our bodies' biological rhythms and physiological sensitivities to the weather and to the diurnal cycle, and plants' similar sensitivities (for example, phototropism, growing season) are closely bound to the cosmos. So, since both plants and human bodies intimately share the affections of the Whole, Epictetus infers that human souls share the affections of the Whole even more so.[43] He argues that (a) the sun can illuminate a huge part of the Whole while leaving dark only the small space that is no larger than what is cast into shadow by the Earth; (b) god causes the sun to revolve, the sun being a small part of god compared with the Whole; therefore, (c) god perceives all things.[44] Epictetus's theological explanations of natural phenomena will seem antiquated and ignorant to a secular scientist. Yet in describing a tightly knit biological community of organisms (plants, humans) interacting in sympathetic conjunction with the other elements of their environment (mountains, rivers, lakes, sun, moon), Epictetus anticipates the notion of an *ecosystem*. This adumbration of the idea of an ecosystem bespeaks the earthy virtue of ecological wisdom, which can and ought to guide our contemporary philosophical synthesis of ecological science with agriculture. Epictetus's nascent notion of an ecosystem thus advances my thesis that we need to implement agricultural practices that harmonize with nature and support living in agreement with nature.[45]

Moreover, Epictetus's teachings deeply influenced Marcus Aurelius. Like Seneca and Epictetus, Marcus also keenly observed nature.[46] In his set of private philosophical reflections traditionally called the *Meditations*, Marcus draws a lesson from arboriculture to demonstrate that people ought to stick together and preserve solidarity.

A branch cut away from the branch beside it is simultaneously cut away from the whole tree. So too a human being separated from another is cut loose from the whole community. The branch is cut off by someone else. But people cut themselves off—through hatred, through rejection—and don't realize that they're cutting themselves off from the whole civic enterprise. Except that we also have a gift, given us by Zeus, who founded this community of ours. We can reattach ourselves and become once more components of the whole. But if the rupture is too often repeated, it makes the severed part hard to reconnect, and to restore. You can see the difference between the branch that's been there since the beginning, remaining on the tree and growing with it, and the one that's been cut off and grafted back. "One trunk, two minds." As the gardeners put it.[47]

Social groups are living wholes of which individual persons are organic interdependent parts. We are myriad cells constituting the single organism of society. This is the Stoic idea of cosmopolitanism. Stoics believe that Zeus established a cosmic community of all rational beings, mortal humans and immortal gods. Rationality transcends differences among racial, ethnic, linguistic, religious, and cultural groups and binds every thinking being as a citizen (*politēs*) of the world (*cosmos*) together into the same, single, all-encompassing city (*cosmopolis*). As a citizen of this cosmopolis, Marcus is responsible for preserving its unity and promoting its harmony. Unfortunately, people too often behave irresponsibly, disregard their neighbors, and sever the ties binding them together. Strife among us is like branches cutting themselves off from their tree. Marcus believes it fortunate that we all have the power to reattach ourselves to our social groups, heal the rifts we have torn in the social body, and return to the fold. The ability to reestablish solidarity—the healing powers of cooperation and collective civic effort toward a common good—is a vital virtue. To work in concert as a single-minded group for the good of the whole, huge organism of which each person is but a small part is a lesson the adept gardener teaches in practicing her art.

Marcus detects beauty in countless subtle features of nature. Observe the wheat we harvest to make flour we bake into bread. Watch closely fig and olive trees.

We should remember that even Nature's inadvertence has its own charm, its own attractiveness. The way loaves of bread split open on top in the oven; the ridges are just by-products of the baking, and yet pleasing, somehow: they rouse our appetite without our knowing why. Or how ripe figs begin to burst. And olives on the point of falling: the shadow of decay gives them a peculiar beauty. Stalks of wheat bending under their own weight. . . . And other things. If you look at them in isolation there's nothing beautiful about them, and yet by supplementing nature they enrich it and draw us in. And anyone with a feeling for nature—a deeper sensitivity—will find it all gives pleasure. Even what seems inadvertent. . . . He'll look calmly at the distinct beauty of old age in men, women, and at the loveliness of children. And other things like that will call out to him constantly—things unnoticed by others. Things seen only by those at home with Nature and its works.[48]

A discerning, sensitive feel for Nature characterizes a naturalist. To recognize the beauty and charm of Nature and be pleased by it characterizes a nature lover. To study the regular, intelligible processes of growth and maturation of fruiting plants is to do horticulture. Those with the keen eyes of a gerontologist notice the fine details of each stage of human life, from childhood to adulthood to senectitude, in women and men. Yet only those of a Stoic bent find beauty in all those fine details and calmly embrace aging as good. Stoics study, learn from, love, and embrace Nature. This enables them to be "at home (*oikos* = *ecos*) with Nature and its works," which is what it is to be *ecologists*.

Marcus learns another lesson from olive trees. Olives on the branch teach us how to live, celebrate nature, and die the right way. "To pass through this brief life as nature demands. To give it up without complaint. Like an olive that ripens and falls. Praising its mother, thanking the tree it grew on."[49] Gratitude to Mother Earth for birthing, nursing, and raising us up is the Stoic's response to Nature. Human beings are olives growing on, and inevitably falling from, the tree that is the natural world. The same insight about change comes from patiently observing grapes. "Grapes. Unripe . . . ripened . . . then raisins. Constant transitions. Not the 'not' but the 'not yet.'"[50]

Another ethical lesson comes from hydrology. "Dig deep; the water—goodness—is down there. And as long as you keep digging, it will keep bubbling up."[51] Persistent effort will succeed in irrigating virtue. This self-confidence is shown when Marcus reminds himself that nothing anyone says or does to him cuts his mind off from clarity, sanity, self-control, or justice. He likens this cognitive purity to being a spring of clear, sweet water that is being cursed by a man who shovels mud or dung into it. The mind of a Stoic is like fresh water that keeps bubbling forth and washing away any pollutants in its stream: "To have that. Not a cistern but a perpetual spring. How? By working to win your freedom. Hour by hour. Through patience, honesty, humility."[52] The inexhaustible plenitude of the bubbling spring inspires Marcus's daily regimen of exercising virtues to win freedom. This is freedom from irritability, deceit, and conceit. Freedom from wickedness comes with cultivating patience, honesty, and humility. Many texts in the *Meditations* liken the never-ending flow of changes in the world to a river.[53]

Repeated observations of fig trees and mindful reflection on how nature operates teach Stoics wisdom. The wise are never surprised. "Remember that it's as shameful to be surprised that a fig tree bears figs as it is that the cosmos produces whatever its *crop* is. And a good doctor isn't surprised when his patients have fevers, or a helmsman when the wind blows against him."[54] Sometimes patients get sicker. Sometimes the wind blows in your face. Only fools are surprised by whatever crops up in the cosmos. Foolishness is a vice we cannot afford to indulge.

Another vice is impatience. In the following text Epictetus lectures a student who believes that he has made such great strides in philosophical wisdom that he fancies himself a big deal and rushes to self-congratulate. Hubris has blinded him to a lesson taught by flowering plants.

Keep your philosophy to yourself for a little while. That is the way fruit comes to be. The seed must be buried and lie hidden for a season and grow incrementally to achieve maturity; but if it produces the ear before the jointed stalk, it never reaches maturity, it's from a garden of Adonis. You too are a kind of plant: you've bloomed too soon and the winter will shrivel you. See what the farmers say about the seeds when the summer heat arrives too soon. They worry that the over-eager seeds will sprout up too lushly, and that a single frost will arrest them and lay bare their weakness. You better watch out too, fella. You've developed with over-eager cockiness, you've jumped up to grab some glory before the time is due. You think you're somebody, fool among fools. You'll be chewed up by the frost, or rather, you've already been chewed up by the frost down at the root, though your stalk still holds a few blooms so you suppose that you're still alive and thriving. Allow us, at least, to ripen as nature wants. Why do you expose us to the elements, why force us? We're not yet ready to withstand the air. Let the root grow, let it next produce its first joint, then the second, then the third. This is how the fruit will naturally force its way out, whether or not I wish it.[55]

Remarkably, Epictetus tries to impart to his impetuous pupil the patience and perseverance of a plant. This is because Epictetus also teaches that god/Nature brought non-rational animals into the world simply to use their sensory stimuli and pursue their bodily desires. Our goal, in contrast, is to live rationally, which includes understanding our sensory stimuli, contemplating Nature, and being appreciative spectators and interpreters of the works of god/Providential Nature.[56] Yet in this text Epictetus cautions his hasty, arrogant student to take his time, internalize the good habits of mind he is still learning, rehearse his lessons slowly, deliberately, and often, effectively sinking deep down the roots of his training (askēsis). Then, little by little, he can send up each joint of his stalk and grow slowly and steadily, safe from the frost. The student who hurries to bloom too soon should emulate the plant paragon.

It makes sense that trees are especially instructive for the Roman Stoics. The ancient Roman scholar Marcus Terentius Varro (116–27 BCE) declared that "the whole of Italy resembles one vast orchard."[57] But the life lessons these Stoics draw from nature are by no means limited to plants. Seneca, Musonius, and Epictetus explicate many exemplary traits and behaviors of nonhuman animals.[58]

Given such a wealth of instructive ethical examples taken from the natural world, why is Musonius, for instance, such a big fan of farming? Why does he think farming is at least as good an occupation as philosophy for becoming good? These questions will be answered by investigating the concept of agrarianism.

The Virtues of Roman Agrarians

Agrarianism is a sociopolitical worldview that views rural society as superior to urban society, and the independent farmer as superior to the paid worker. Agrarians believe that farming as a way of life shapes ideal social values and corrects false beliefs about the good life held by urbanites. The agrarian ideal is epitomized by the ancient Roman legend of Cincinnatus.

Lucius Quinctius Cincinnatus (c.519–c.430 BCE) was a patrician and suffect consul in 460 BCE. According to legend, in 458 BCE a Roman army was besieged by invaders. Cincinnatus, an exiled commander around sixty years old, was working his small farm when he was summoned by the Roman people to rescue the state. Within sixteen days he assembled an army, defeated the invaders, surrendered his power, and returned to the plow. In the Roman tradition, the story of Cincinnatus was a shining example of superlative leadership, civic dedication, austere modesty, and pastoral virtue. His story was so celebrated during the American Revolution that a mural was painted in the U.S. Capitol building to commemorate it and instruct lawgivers and citizens in civic ideals.

A second exemplary agrarian was the senator, orator, and historian Marcus Porcius Cato (Cato the Elder) (234–149 BCE). Born to an ancient plebeian family of farmer-soldiers, he led a distinguished military career and wrote *De Agri Cultura*, a farmer's notebook, c.160 BCE.[59] He was revered for his practicality, austerity, and efforts to stem the tide of extravagance and licentiousness. Heralded for his natural sagacity and many civic and political accomplishments, in old age he was often called Cato the Wise.

Cincinnatus and Cato the Elder were stoics with a small "s," but were no fools, as Stan the patient was. Cato the Elder's great-grandson, Marcus Porcius Cato Uticensis (Cato the Younger), was a declared Stoic philosopher with a capital "S." The great-grandson emulated his great-grandfather. Cato the Younger was heroized by Seneca for his tenacity, faultless integrity of character, and valiant opposition to the tyranny of Julius Caesar.

Cato the Elder is the main character of Cicero's dialogue *Cato the Elder on Old Age*. The other two characters of this dialogue, nearly fifty years younger than Cato, are Scipio Aemilianus (185–129 BCE) and Gaius Laelius (c.186–? BCE). Scipio, a prominent patron of Greek and Roman writers and philosophers, had an illustrious military career and was judged by the historian Polybius to be one of the purest and noblest men in history. Laelius, Scipio's old army buddy, also won military renown. But Laelius, who studied with the capital "S" Stoics Diogenes of Babylon and Panaetius, gained great philosophical learning. Admired for his erudition, equanimity, integrity, justice, and nobility of thought, Laelius earned the title "the Wise." Thus, whereas Scipio Aemilianus was a small "s" stoic, Laelius the Wise was a capital "S" Stoic.

In Cicero's dialogue, Scipio and Laelius marvel at Cato's surpassing wisdom and the fact that old age never burdens him. Cato attributes his wisdom to following nature as the best guide and obeying her as a god. Nature, he avers, is not so careless a playwright as to plan out each act of life's drama while neglecting the final act. He compares the necessity of the final act to the way orchard fruits and crops of grain ripen over time, gradually shrivel, and get ready to fall. Cato calmly accepts his mortality as a limit nature sets. Laelius and Scipio urge him to pass on to them the principles that allowed him to reach his destination free of worry. In reply, Cato delivers a long panegyric of agriculture. The delights of agriculture he judges best suited to the life of a wise old man. Mother Earth he likens to a bank that reliably returns generous interest on the principal. He praises the soil's power to embrace, warm, and transform scattered grains of wheat into gradually rising stalks of sheathed ears in ordered rows. Cato revels in "the inherent force of all those things which are generated from the earth—a force that, from the tiny fig-seed, or grape-stone, or from the smallest seeds of other fruits and plants, can produce such mighty trunks and boughs." He says the farmer finds joy not only in his cornfields, vineyards, orchard, and garden, but also in his meadows, woodlands, and the bees swarming among an infinite variety of flowers. All the farmer's tasks yield crops alluring to the eye and delicious in taste. Thus, Cato gives three reasons to think that no life can be happier than that of the farmer. First, the aesthetic charms of every aspect and element of the farm are overt. Second, the duty the farmer performs sustains and benefits all of humanity. Third, the plenty and abundance of the farm in providing everything that tends to nurture humanity engenders worship of the gods, presumably as gratitude for their providential care of human beings. He concludes: "Nothing can be more abounding in usefulness or more attractive in appearance than a well-tilled farm, and to its enjoyment old age not merely offers no obstacle, but even entices and allures." Nowhere else can an old man bask in cordial sunshine, or shake off winter's chill beside a cozy fire, or in the summertime cool himself under shady trees by burbling streams. Compared to the joys of agriculture, horses, weapons of war, ball games, swimming contests, foot races, and idle games of chance so popular with city folk fail to attract Cato. He delights in the bounty and beauty of a farm while embracing the limits nature imposes, including mortality.[60]

One commentator remarks on how the untidy habit of the vine "to sprawl is arrested, tamed, pruned and moulded by the loving skill of the vine-dresser—a glowing tribute to man's *capacity to tame nature* in the interest of civilized living."[61] Later, I will show that a key concern of the Midwest Stoic in evaluating agricultural methods must always be whether and to what extent *taming* nature harmonizes with living *in agreement* with nature. All farming inevitably disturbs land and water a little or a lot. Thus, determining which practices are sustainable and prudent in the long run requires circumspect judgments grounded in virtues.

Cicero declares that "there is no kind of gainful employment that is better, more fruitful, more pleasant, and more worthy of a free man than agriculture."[62] He praises agrarian life as the teacher of parsimony, industry, and justice.[63] Thus Cato the Elder, Cicero, and the Stoic Musonius Rufus all extol agrarian virtues, which include self-sufficiency, diligence, hardiness, parsimony, perseverance, equanimity, modesty, justice, and civic responsibility. Farmers promote the common good by producing food for everyone. Recall the lesson from Marcus about what we can learn from arboriculture. Agrarian virtues animate a philosophical perspective and way of life I dub Midwest Stoicism.

From Roman Stoic Agrarianism to Midwest Stoicism

A farmer's fields, orchard, vineyard, garden, and woodlands are like an artist's canvas, easel, pigments, pastels, and oils. A farmer's hoe, plow, rake, fork, scythe, bolo, spade, axe, and mattock resemble an artist's brushes, knife, and sponge. Born near the tiny town of Anamosa in Jones County, Iowa, the artist Grant Wood (1891–1942) grew up in Cedar Rapids. He studied at the Handicraft Guild in Minneapolis and the School of the Art Institute of Chicago. In the 1920s in Europe, he studied impressionism and post-impressionism, and gained a strong appreciation for the realism of Jan van Eyck. Wood became the leading proponent of the American movement of Regionalism in the arts, advancing figurative painting of rural Midwestern themes. His childhood country school is depicted on the 2004 Iowa State Quarter. He is buried at Riverside Cemetery in Anamosa. The Figge Art Museum in Davenport displays some of Wood's personal effects and works of art. No artist is more quintessentially Midwestern.

Wood's most emblematic work, *American Gothic*, remains one of art history's biggest icons. "Featuring a stoic portrait of a farmer and his daughter, this painting offers a fascinating glimpse into life in the rural United States."[64] The house in the portrait was built in a Gothic Revival style called Carpenter Gothic. The characteristically "rural" appearance of this abode inspired Wood to imagine "American Gothic people with their faces stretched out long to go with this American Gothic house."[65] For his models, Wood chose his younger sister, Nan Wood Graham, and his dentist, Dr. Byron McKeeby.

> Wood dressed the figures in clothing typical of a farming family. Wood Graham, for example, wears a colonial print apron and has a cameo pendant pinned to her high-collar, while McKeeby wears overalls and carries a pitchfork. He also opted to give the figures stoic expressions—a choice that many Iowans misinterpreted as an attempt to portray them as "pinched, grim-faced, puritanical Bible-thumpers."

Wood, however, stressed his appreciation for his home state and stated that this was not the case.[66]

Iowa farmers are not joyless religious zealots. They just look like small "s" stoics. The faces of *American Gothic* are those of Midwest stoic agrarians.[67] How closely do these expressions resemble those of Dr. Gracey's stony farmer? The evidence furnished by Wood and Gracey indicates that small "s" stoics are common in the Midwest. Do these folks who *look* stoical *think* like big "S" Stoics?

Recall from the opening section the Great Eight Midwestern states. Four are contiguous with Kentucky. Given its geography and history, the northern tier of Kentucky along the Ohio River can be regarded as at least fuzzily Midwestern. This area includes Henry County, home of the acclaimed farmer, author, cultural critic, and environmental activist Wendell Berry. Berry quotes from the fourth of Virgil's *Georgics* to tap into the ancient Roman stoic agrarian theme of the small farmer leading an abundant life on a discarded scrap of land.

> I saw a man,
> An old Cilician, who occupied
> An acre or two of land that no one wanted,
> A patch not worth the plowing, unrewarding
> For flocks, unfit for vineyards; he however
> By planting here and there among the scrub
> Cabbages or white lilies and verbena
> And flimsy poppies, fancied himself a king
> In wealth, and coming home late in the evening
> Loaded his board with unbought delicacies.[68]

Berry tracks the folk tradition of this old squatter from ancient Rome to today. He neither has nor needs much land. His land is often marginal. He always associates frugality with abundance. Through this agrarian lens, Berry appraises the value of a small scrap of land by reference to having no land at all. Agrarians know that to be landless is to be existentially lost. Consequently, the old subsistence farmer in Virgil's poem is both wise and happy to accept "an acre or two of land that no one wanted." "If you have no land you have nothing: no food, no shelter, no warmth, no freedom, no life."[69] Berry emphasizes that respect for limits is essential to agrarianism.

Agrarian farmers see, accept, and live within their limits. They understand and agree to the proposition that there is "this much and no more." Everything that happens on an agrarian farm is determined or conditioned by the understanding that there is only so much land, so much water in the cistern, so much hay in the barn, so much corn in the crib, so much firewood in the shed, so much food in the cellar or freezer, so much strength in the back and arms—and no more. This is the understanding that induces thrift, family coherence, neighborliness, local economies. Within accepted limits, these virtues become necessities. The agrarian sense of abundance comes from the experienced possibility of frugality and renewal within limits.[70]

Berry's agrarians accept and abide by limits just as Cato the Wise accepts that he is playing out the final act of his mortal life, a limit set by nature. Cato beams about the bounty and beauty of his farm as Berry invokes Virgil's old Cilician fancying himself a king feasting on cabbages. The thrift, frugality, and neighborliness of Berry's agrarians reflect Cicero's lessons in parsimony, industry, and justice learned from farming.

Seneca, Musonius Rufus, Epictetus, and Marcus Aurelius also honor limits. Seneca shows how reverence of nature requires respecting limits. Admiring the beauty of nature reveals the world's orderliness and pulls us away from fear, anger, greed, petty indulgences, and carnal vices. Consciousness of the vast grandeur of the cosmos shrinks our self-absorbed worries to vanishing points. Like Berry, Musonius sees the virtue of freedom realized by farming. Musonius believes the benefits of farming include abundant leisure for contemplation, living outdoors in the country, and procuring one's own necessities. Farming fosters the virtues of industriousness, seemliness, justice, and courage. But for Musonius the acceptance of limits is particularly salient in the agrarian virtues of austerity, self-reliance, and self-control. Constant vigilance about the limits of our powers, about what is and is not up to us, is fundamental to Epictetus's teaching. The agrarian sense of abundance grown from the lived possibility of frugality and renewal within limits described by Berry resonates strongly with Epictetus's insistence that by teaching us perseverance, patience, and wisdom, nature empowers us to improve ourselves, thereby granting us abundant opportunity to flourish. Marcus Aurelius believes that from organic bodies we can learn social cohesion and mutual cooperation—what Berry calls neighborliness. Like Seneca, Marcus too summons sensitivity for the beauty of nature. But the lesson of limitation he so often rehearses is that nature teaches us that transience and change are certain. To judge this certainty as good fosters feeling and being at home with nature and its works. Marcus reminds himself that observing nature teaches us patience, how to see the big picture, comprehend the enormity of time, and so never be surprised or upset. Indeed, all the Roman Stoics emphasize the limit of mortality. No philosophers take limits more seriously than Stoics.

Along with the ecological limits Berry describes, biology imposes others: not only mortality, but fragilities like susceptibility to disease, vulnerability to injury, and senescent decrepitude. Stoics directly and habitually confront all these limits. You're worried, anxious, afraid? Is it about something that is up to you to control? If not, don't sweat it. Don't waste your time or effort wanting to change what isn't yours to change. What you're able to do is limited. Each day you need water, food, and rest. You're mortal. Just like all other people, animals, and plants now alive, your death is certain. Just like all those people, animals, and plants that came before you, they turned to dust and are gone. You only have so many days and no more. You only have so much energy and no more. You only have so much time to complete your tasks and no more. Stop dragging your feet. If you want to become a better person, you must get to it *now*. No more delays. No more excuses. Welcome what you're given as a temporary gift. Make the best of what you've got right now—*that* is up to you. If you want to be happy, be wise. To cherish what you have is wise. To let go, without complaint, of what was only loaned to you is wise. Frugality and renewal within limits resonate in Midwest Stoicism.

Now, a worry could be raised about how specific Stoic agrarianism is to the Midwest. Does Berry's agrarian ideal apply to farming in any region of the United States, whether New England, the South, California, or the Pacific Northwest?

First, in various ways the history of each region subtly colors its cultural tendencies and norms. The slavery upon which Southern plantations of the nineteenth century operated, for instance, leaves a deep and lasting stain of social and ecological injustice in the South. This history of racial injustice in the states of the Confederacy does not extend to agrarians in the Union who fought for abolition. One might also suggest that the suffering and hardship of the Dust Bowl of the 1930s forged a stoic endurance and resoluteness of agrarians across the Midwest. Few Midwest agrarians lease their fields for oil and natural gas extraction compared to agrarians in Oklahoma and Texas.[71]

Second, the filter of cultural expectations can obscure the gauging of (agrarian) values, virtues, and vices. It may be tougher for Midwesterners to detect the virtues of courtesy, patience, generosity, modesty, and friendliness in East Coasters than in fellow Midwesterners. A stereotypical Californian, on the other hand, may strike a Midwesterner as too self-entitled, too self-absorbed, too fragile to handle harsh weather, and too fond of automobiles. So, perhaps what varies from region to region is more the styles in which certain virtues are expressed rather than their actual distribution. Uncouth behaviors of members of one region may be off-putting to those of another without necessarily being reliably objective measures of vice.

What of agrarianism in the South? Is Wendell Berry, a Kentuckian, a dubious choice as a proponent of Midwestern Stoic agrarian values? I do not claim that the Stoic agrarianism that resonates with elements of Berry's philosophy takes root exclusively in the Midwest. Rather, I opine that Stoic agrarian virtues are in high relief in the Midwest while admittedly cropping up in other locales in North America and beyond.

Today very few Americans and very few Midwesterners are farmers. The majority of those involved in agriculture belong to Big Ag—the huge corporations that control the nation's (and increasingly the world's) land, seeds, plants, food, and farm machinery. These workers are not landed because they are either precariously employed farmhands with no claim to the land they work, or they are agribusiness employees, or they are corporate shareholders with no real connection to the land. The very few who are independent farmers rely heavily on farm subsidies and agribusiness contracts. Consequently, the millions of Midwesterners living in cities, suburbs, and rural towns conquered by the profit-ravenous, labor-exploiting, and ruthlessly environment-damaging behemoth of Big Ag cannot plausibly be said to embody Stoic agrarianism. That is precisely why it is urgent for Midwesterners, indeed all Americans, to transform their lives. I will return to this point in the final section of this chapter.

What of Stoic theology? Must Midwest Stoics believe in divine Providence? For a post-Darwinian, belief in evolution instead of divine Providence does not preclude the belief that the universe is vastly larger and grander than we are, that all of its parts interconnect, and that we can and should exercise our intellects to live in harmony with this huge cosmic whole of which we are tiny parts. A Midwest Stoic need not agree with Epictetus's belief that the Stoics' Zeus commands plants to flower and drop their fruits in order to affirm the idea of an ecosystem, admire nature's beauty, and glean from nature lessons in frugality, self-control, and modesty.

One could object that farmers hold no monopoly on such virtues. According to agrarianism, farming is better than other occupations. Why? Compare being a salesperson, commodity trader, or proprietor of a brothel. Couldn't trading commodities and services as "preferred indifferents" in a free market count as admirable in capitalism? The brothel, one could argue, provides sexual services to cater to a demand for erotic "preferred indifferents." The profitability of the brothel could result largely from the proprietor's virtues of industriousness, frugality, and entrepreneurial resourcefulness.

A problem with this argument is that any occupation motivated by vice is objectionable. Is the motive for buyers and sellers greed? Does the monetization of sexual desire feed the greed of pimps and sex workers while fueling the lust of their clients? To work a job motivated by greed, lust, envy, fear, or anger is to suffer from vice. Stoics regard greed, fear, anger, and similar passions as violent, pathological disturbances of sick minds. The farmer who mistreats his employees, exploits and abuses animals, cheats on his taxes, pollutes water and air, bribes officials, and is driven by greed is just as reprehensible as the brothel owner afflicted by the same vices. So it's true that agrarians have no monopoly on the specified agrarian virtues. But Midwest Stoic agrarians who are virtuous are ipso facto superior to those who aren't.[72]

Sometimes vices are disguised as virtues. As Judith Lee rightly observes, some regard the Midwest not as the Heartland of virtues but rather as a Hinterland populated by rubes and suburbanites, a backwater so inhospitable that its native-born artists and intellectuals flee to the coasts. Nonetheless, Lee believes that "Americans associate the pastoralism of farm life with cultural wholesomeness and thereby identify the Midwest as 'genuine America,' 'the keeper of the nation's values.'"[73] By identifying the Midwest as "genuine America," Lee expresses the perspective of what I will term a Midwest Stoic *agrarianist*. As we have seen in the Roman agrarians, appreciation of nature's limits, beauty, and bounty, and the virtues of parsimony, assiduity, and justice, are in fact transnational, cosmopolitan values that Midwest Stoic *agrarians* embrace (see the later discussion for the distinction between *agrarianists* and *agrarians*).

Midwest Stoic Agrarianism and Environmental Virtue Ethics

I have been arguing that Midwest Stoicism can enable us to effectively cope with and ameliorate today's environmental problems. Midwest Stoicism calls for us to live in agreement with nature, which means especially living in agreement with reason, the perfection of which is virtue. Applying the matrix of virtues animating Midwest Stoicism to ethical reflection on contemporary agriculture and the environment yields the following thesis. Possessed of the virtues of ecological wisdom, frugality, simplicity, modesty, self-control, patience, perseverance, courage, assiduity, cooperation, justice, appreciation of beauty, and respect of limits discussed above, Midwest Stoic agrarians (MSAs) embrace as sustainable those agricultural practices that harmonize with nature by living sustainably within the limits inherent to ecosystems. Conversely, MSAs oppose agricultural practices manifesting vices of selfishness, arrogance, vanity, greed, self-indulgence, laziness, wastefulness, injustice, cruelty, or exploitation. Agricultural practices contrary to nature undermine human flourishing and characterize grave ecological folly. Consequently, these vicious practices must end.

Environmental abuses inflicted on the Midwest have left deep, wide, and lasting scars. The outlook of experts is bleak.

> Often well-intentioned exploitation of midwestern bounty has left a dismal ecological legacy. Human population surges erased some animals and many more plants from the landscape. Rapacious logging altered environmental balance. The plow literally destroyed the prairies. The wind and water erosion that followed stripped away centuries of topsoil. Fertilizers and mining byproducts and industrial waste polluted waterways and poisoned the land. Dams and powerplants diminished the quality of water and air. Today midwestern states battle against histories of offense and neglect.[74]

The ethical imperative of halting this devastating spiral of ecological folly has never been more urgent.

The ramifications of today's ecological challenges facing Midwestern farming communities, consumers, and citizens are ethical, economic, and political. Today less than 2 percent of Midwesterners are still farmers. "The work ethic that trademarked the American agrarian lifestyle has been replaced by agrotechnology."[75] Consider corn. The vast majority grown in the Midwest is not sweet corn eaten by people but is either feed corn for livestock, converted into ethanol, or transformed into high fructose corn syrup.[76] The dominance of this monoculture spawned the slogan "Corn is King." If so, King Corn is a cruel tyrant, because "abundant, cheap midwestern corn makes the cheap, supersized soft-drinks and hamburgers that have expanded waistlines wherever the American diet prevails."[77] In a bizarre circle, fossil fuels are made into fertilizer to grow corn to make fuel for cars and trucks. "Chemical fertilizers get triply blamed: in addition to polluting waterways, they contribute to global warming and energy dependence, through their reliance on oil."[78] Throughout the Midwest a cohort of multinational agribusinesses (Big Ag) dictates what is grown (King Corn), how it is grown (using petrochemical fertilizers, GMOs, pesticides, herbicides), and the costly externalities produced (pollutants).

The political power of Big Ag is immense. Big Ag demands and receives huge, perpetual, ballooning federal subsidies from taxpayers. Moreover, Big Ag is guilty of gagging whistleblowers, horrific abuse of animals, and rampant environmental contamination.[79] These ecological harms are externalized by the meat and dairy industrial complexes. Concentrated Animal Feeding Operations (CAFOs) subject their workers to harsh working conditions with poor protections. The hegemony of Big Ag has driven small, independent farms to near extinction. The relentless, crushing economic pressures on small farms directly increases psychological stress on farming families (recall the illness of South Dakota farmer Stan in an earlier section), which in turn causes higher rates of suicide among farmers.[80] Former President Trump's tariffs made matters worse, driving up bankruptcies and suicides among farmers.[81]

The detriments of the cattle industry are old news,[82] though poultry and hog CAFOs and processing facilities are in many ways even worse for the animals, the workers, and the environment.

> Nitrogen and phosphorus in runoff from dairy farms and associated feedlots augment the growth of algae in ponds and boost nitrate levels in drinking water. Such runoff poses health risks to local populations. In Milwaukee in 1993, drinking water contaminated with cryptosporidium from livestock manure killed sixty-nine people and more than 400,000 had to be treated for various symptoms.[83]

Industrial agrobusinesses are to blame for mercury accumulation making fish in Midwestern lakes unsafe to eat. That the meat and dairy industrial complexes endanger public health is endemic to their business design. Profits are enormous because the manure and toxic pollutants generated by privately owned CAFOs become externalities the public is forced to pay for. CAFOs pump feces, urine, antibiotics, and a range of poisons into giant holding lagoons. These lagoons release their pollutants into groundwater, streams, and rivers. No CAFOs are required to treat any of the toxic rivers of animal waste they discharge. CAFOs also emit staggering quantities of carbon dioxide, methane, and other greenhouse gases, polluting the air and worsening anthropogenic global climate change.[84] The environmental harms of Big Ag include dependence on fossil fuels; soil erosion; freshwater depletion; intensive use of pesticides, herbicides, and petrochemical fertilizers; loss of biodiversity; greenhouse gas emissions; air pollution; water pollution; and zoonotic diseases. Though not originating in the Midwest, the latest zoonotic disease spawned by the meat industry is COVID-19.[85]

The effects of anthropogenic climate change are felt throughout the Midwest. Storms, flooding, and drought grow in severity and frequency. The prevalence of pests, disease, and competition from non-native species increases. The rate of warming in the Midwest has markedly accelerated over the past few decades. "Projected changes in precipitation, coupled with rising extreme temperatures before mid-century, will reduce Midwest agricultural productivity to levels of the 1980s without major technological advances."[86] Global weirding, destructive weather events, and loss of property inflate insurance claims, which in turn ratchet up insurance premiums for those who can still afford insurance.[87] The U.S. Global Change Research Program reports:

> In the absence of more significant global mitigation efforts, climate change is projected to impose substantial damages on the U.S. economy, human health, and the environment. Under scenarios with high emissions and limited or no adaptation, annual losses in some sectors are estimated to grow to hundreds of billions of dollars by the end of the century. It is very likely that some physical and ecological impacts will be irreversible for thousands of years, while others will be permanent.[88]

How can Stoicism address these daunting challenges? I have argued that Stoics are naturalists—really, ecologists—who idealize the virtues of self-sufficiency, diligence, hardiness, perseverance, equanimity, simplicity, modesty, civic responsibility, cooperation, and solidarity. Stoic agrarians admire the diligence and solidarity of bees working for the common good of their hive.[89] In antiquity "the social life of the bee caused it to be a symbol of ideal political structure and its industriousness was constantly praised."[90] MSAs embody these virtues.

Critics of Midwest Stoic agrarianism challenge this claim. They deny that Midwesterners are moderate, thoughtful, persistent, or modest. To them, we are hinterlanders, bumpkins. To these critics, "Midwestern moderation and thoughtfulness become blandness and timidity; their persistence, pig-headedness; their modesty, banality."[91] Stoics would reply that recognizing the limits of one's knowledge, as Socrates did, should prompt a person to be modest. Stalwart preservation of integrity of character is not pig-headedness. Thoughtfulness is indispensable for gaining wisdom. Given the disastrous threats we face, ecological wisdom demands bold action rather than the least shred of timid complacency.

Midwest Stoic agrarianism readily lends itself to a *practical* environmental virtue ethics.[92] The practice of farming in the right way is just one part of living virtuously, sustainably, and well within an ecosystem. Living well within an ecosystem means that all of us, farmers and nonfarmers, must live a certain way. Everybody has their role to play. This cosmopolitan idea is stressed by the ancient Stoics. Seneca emphasizes that living well begins with studying nature, learning from it, and not just acknowledging its beauty but holding that beauty in awe. Musonius Rufus stresses that farming teaches industriousness, hardiness, seemliness, self-reliance, self-control (indeed vegetarianism), courage, and freedom. Epictetus imparts the lesson that nature teaches us patience, fearlessness, and perseverance, as well as our connectedness to all of nature. When properly nourished by nature in mind and body, we gain the power to flourish, according to Epictetus. Marcus Aurelius emphasizes that nature teaches us our interdependence and embeddedness in the cosmic scheme. His ethic of cosmopolitanism underscores the urgent need to cooperate and to build and preserve social solidarity if we are to do our parts and live well. Marcus's recognition that we utterly depend on Mother Earth, that we belong to Earth and not vice versa, is fundamental to MSA. The Roman agrarians also contributed to the holistic environmental virtue ethics articulated in MSA. The rigors of farming taught Cincinnatus civic dedication and selfless leadership. Cato the Elder learned from agriculture the wisdom of embracing the limits nature imposes while delighting in a farm's beauty and bounty. Cicero concurs, noting that the agrarian life instills parsimony, industry, and justice. These many virtues of the ancient Roman agrarians were inherited by Grant Wood's friends and neighbors. The sturdy, stoical handsomeness of Midwestern agrarians is memorialized in his *American Gothic*. Wendell Berry honors this ancient tradition of agrarianism for understanding and accepting the inescapable limits that induce thrift, family coherence, and neighborliness. Midwest Stoicism can in fact be considered the philosophy explicitly about how to thrive amidst inescapable limits.

Thus construed, MSA provides considerable conceptual tools for responding to the serious environmental and sociopolitical challenges facing Midwesterners. For example, the Midwest Stoic virtues of simplicity, frugality, and temperance call for sustainably grown, locally sourced grains, vegetables, fruits, legumes, nuts, and seeds for human beings to eat. According to MSA there is no wisdom in cycling countless tons of GMO feed corn through millions of miserable ungulates, hogs, and hens crowded into CAFOs when human beings can eat plants directly. The fact that whenever temperatures rise above freezing motorists traveling through the rural Midwest are often assaulted by the godawful stench of thousands of confined bovines signals that CAFOs emblematize a food system grievously contrary to nature. The ethics of Midwest Stoic agrarians diagnose the vices endemic to Big Ag. These include greed, wastefulness, secrecy, deception, exploitation, violence, environmental injustice, and marketing gustatory decadence to unwitting Midwesterners.

Which practical steps address this vicious system of food production and the diseases stemming from both the production and consumption of factory-farmed meat and dairy products? A first, obvious step is a more rapid, more widescale shift to locally sourced, plant-based foods. This shift cannot take place only in the Midwest. It must be widescale. This requires a broad, indeed global, perspective. Stoicism provides exactly this cosmopolitan outlook. The meat and dairy industrial complexes are far too wasteful of inputs (fossil fuels, topsoil, clean water, feed corn, etc.), far too destructive of ecosystems, and far too polluting of water and air. Big Ag belches out far too much and far too many greenhouse gases, destabilizes ecosystems, and recklessly endangers human health to reap staggering profits. These are big problems that must be grasped from a big-picture, cosmopolitan perspective.

Must Midwesterners be or become landed to enact Midwest Stoic agrarianism? No. Most Americans will never own more than one third of an acre. But a citizen need never become landed to cultivate Midwest Stoic agrarian virtues in selecting and obtaining food. Everyone must eat. So, everyone should strive to eat responsibly. Nearly everyone buys food. So, everyone who buys food should strive to buy responsibly.[93]

A second obvious remedy is environmental activism. Here MSA recalls Cato the Younger, who opposed the tyranny of Julius Caesar. Today Midwesterners can draw inspiration from the capital "S" Stoicism of Cato the Younger to oppose the tyrannies of corrupt politicians, extreme economic inequality, science deniers, Big Ag, and systems of environmental injustice, racism, sexism, and bigotry. We ought to mobilize to promote measures to make voting easier and to safeguard free and fair elections. We ought to oppose unjust practices like gerrymandering. We can work to persuade others of the wisdom of abolishing the Electoral College and remaking the deplorably undemocratic body known as the U.S. Senate. It's patently unfair that a voter in Wyoming wields far greater clout than a voter in California or New York. Each adult citizen of the United States, the District of Columbia, and the adult residents of its Territories deserve to have the same equally weighted vote for representatives in Congress and federal elections. Egalitarian justice demands it. We must work harder to make our democracy better, that is, more democratic.

But isn't agrarianism elitist? It views rural society as superior to urban society and the independent farmer as superior to the paid worker. Here I redefine this as the belief of *agrarianists*. Presumably agrarianists are gleeful that each Wyoming farmer-voter exercises greater voting power than one urbanite in New York City plus a second in Los Angeles plus a third in Chicago added together.[94] Stoic *agrarians*, in contrast to agrarianists, subscribe to the (less contentious?) belief that farming as a way of life can teach healthier social values and can correct false beliefs about the good life (held by non-Stoics). One false belief is that higher consumption of energy and resources at the cost of environmental degradation makes one's life better. Another is that owning things makes one's life better than making things. A third false belief is that quick gratification beats slow progress. One doesn't have to be an urbanite to succumb to consumerism, hedonism, and insidious forms of competition. But crowded (and yet lonely) urban living can often fuel selfishness and aggressive behavior more than living in a smaller, more closely knit community. So, Stoic agrarians champion real egalitarianism among all citizens. Urbanites who attain ecological wisdom (including parsimony, industriousness, simplicity, patience, thrift, temperance, fairness, living within limits, sustainability, familial coherence, and neighborliness) can live in agreement with and admire nature just as much as rural Stoic folks with the same virtues.

A third measure for Stoic agrarians is reproductive restraint for the most consumptive. Americans make up 4 percent of the global population but consume a quarter of the natural resources and a third of the planet's energy. Each generation hands down to the next this pernicious lifestyle of excessive consumerism. Sustainability applied to family planning calls for less procreating and more adopting. Consequently, the habit of greedily gobbling material goods makes having fewer children an urgent responsibility for the affluent in developed countries. Our generation owes it to future generations.[95] Stoic agrarians, fortified by an environmental virtue ethic that champions sustainability, food sovereignty, and grassroots organization to tackle the greatest collective-action problem humanity has ever seen,[96] offers a hopeful vision. Learning to live agreeably with nature is imperative if we are to make the best of the Anthropocene. Americans must lead, not follow, in this global challenge. We must all pull together like busy little bees to save this, our common, and only, hive.[97]

Acknowledgments

I thank Randy Feezell, Scott Aikin, Tom Blackson, Ian Smith, Matt Ferkany, Jill Dieterle, Wade Tornquist, and an anonymous reviewer for comments that improved this essay.

NOTES

1. This disambiguation corrects a common confusion about being a Stoic in contrast to being stoical.

2. See J. W. Slade and J. Y. Lee, eds., *The Midwest: The Greenwood Encyclopedia of American Regional Cultures* (Westport, CT: Greenwood Press, 2004), xxi. Judith Yaross Lee writes: "The region's identity crisis also owes something to disagreement over exactly what areas constitute it. In fact, the region lacked an agreed-upon name until after the Civil War, perhaps in a rejection of the conflict's north-south orientation as westward expansion reached the Pacific coast."

3. Slade and Lee, *The Midwest*, xxi.

4. The editors of *The Midwest: The Greenwood Encyclopedia of American Regional Cultures* judge all and only the Great Eight states to constitute the Midwest.

5. Slade and Lee, *The Midwest*, xxi.

6. *US Census Bureau, Position of the Geographic Center of Area, Mean and Median Centers of Population: 2020, Census.gov.*

7. Originally so designated by historian Walter Havighurst; Slade and Lee, *The Midwest*, xxi.

8. A. A. Long and D. N. Sedley, *The Hellenistic Philosophers* (Cambridge: Cambridge University Press, 1987), 394–398.

9. William O. Stephens, "Stoic Naturalism, Rationalism, and Ecology," *Environmental Ethics* 16, no. 3 (1994): 275–286.

10. Contemporary Stoics in this camp include the late Lawrence Becker, Massimo Pigliucci, and Piotr Stankiewicz.

11. Nigel Glassborow, Kai Whiting, and Leonidas Konstantakos are examples.

12. Ancient Stoics understood "physics" to be the study of nature (*phusis*), including all physical reality, including what we today would regard as both ontology and theology.

13. "Logic" for the ancient Stoics was the study of *logos*, which included principles of reasoning and argument, which we today would call logic, but also rhetoric, the study of parts of speech, and the philosophy of language, including the ontology of propositions.

14. The following accounts of Stoic ethics and physics borrow from my "The Stoics and their Philosophical System," in *The Routledge Handbook of Hellenistic Philosophy*, ed. Kelly Arenson (New York: Routledge, 2020).

15. How, exactly, should we imagine the logic seeds doing these things, according to the Stoics? To be clear, the Stoics offered nothing approaching a granular causal explanation of these processes. We contemporaries would want such a granular explanation articulated in our current scientific terms in biochemistry, genetics, etc. The ancient Stoics had no such nomenclature. So, here I merely try to flesh out, in an admittedly speculative way, the Stoic theory in conceptual terms familiar to us.

16. Douglas R. Gracey, "The Case of the Stoic Farmer," *Heart & Lung* 2 (1973): 744–746.

17. Gracey, "The Case," 745.

18. Though attributed to Alan Watts, this fable may be much older.

19. Yu Liu, "The Complexities of a Stoic Breakthrough: Matteo Ricci's *Ershiwu Yan* (Twenty-five Paragraphs)," *Journal of World History* 24 (2013): 823–847; Antonia Macaro, *More Than Happiness: Buddhist and Stoic Wisdom for a Sceptical Age* (London: Icon Books, 2018). Stoicism and Daoism are compared in Jiyuan Yu, "Living with Nature: Stoicism and Daoism," *History of Philosophy Quarterly* 25 (2008): 1–19.

20. Also known as Seneca the philosopher or Seneca the Younger, to distinguish him from his less illustrious father the rhetorician.

21. In antiquity, ordinary folks often regarded these phenomena as threatening signs of divine displeasure, a view of divinity rejected by Stoics.

22. Seneca, *Natural Questions*, trans. H. M. Hine (Chicago: University of Chicago Press, 2010), 13.

23. Seneca, *Natural Questions*, 14.

24. Seneca, *Natural Questions*, vii 30.3, 133–134.

25. Seneca, *Natural Questions*, vi 32.1, 112.

26. Seneca, *Natural Questions*, ii 59.2, 191.

27. Gareth D. Williams, *The Cosmic Viewpoint: A Study of Seneca's* Natural Questions (New York: Oxford University Press, 2012), 9, 11.

28. Seneca, *Letters on Ethics*, trans. M. Graver and A. A. Long (Chicago: University of Chicago Press, 2015), *Letter* 73.15, 228.

29. Seneca, *Letters on Ethics*, *Letter* 95.12, 369; *Letter* 95.59, 378.

30. Musonius Rufus, *Lectures and Sayings*, ed. William Irvine, trans. C. King (self-pub., CreateSpace Independent Publishing Platform, 2010), *Lecture* 11. All quotations from Musonius are from this source. Musonius Rufus, *Lecture* 11.1.

31. Musonius Rufus, *Lecture* 11.2, 51.

32. Musonius Rufus, *Lecture* 11.2.

33. Musonius Rufus, *Lecture* 11.3.

34. Musonius Rufus, *Lecture* 18a and 18b. See William O. Stephens, "Stoicism and Food," in *Encyclopedia of Food and Agricultural Ethics*, ed. D. M. Kaplan (Dordrecht: Springer, 2018).

35. This belief is a key tenet of agrarianism.

36. Musonius Rufus, *Lecture* 11.4.

37. Musonius Rufus, *Lecture* 11.5, 52.

38. Musonius Rufus, *Lecture* 11.6.

39. Epictetus, *Discourses, Fragments, Handbook*, trans. Robin Hard (Oxford: Oxford University Press, 2014), *Disc.* 1.19.12; cf. 1.22.16.

40. Epictetus, *Disc.* 1.4.30–32; *Disc.* 2.23.5.

41. Epictetus, *Disc.* 2.8.12–18; 1.14.12; 1.14.6; 1.30.1; 2.14.11.

42. Epictetus, *Disc.* 1.14.1–3.

43. Epictetus, *Disc.* 1.14.4–5.

44. Epictetus, *Disc.* 1.14.10.

45. We can and perhaps should reject the particular Stoic view that all that happens in nature is for the best. Yet we can and should accept a Stoic-like claim that living harmoniously within an ecosystem is good for us. If we accepted this, we would adopt a different set of agricultural practices. As part of adopting these agricultural practices,

we would have to adopt a different set of behaviors as virtues. These virtues would not be identical to Stoic virtues, but they would resemble them in certain important ways. Our focus would not be to indulge our own pleasures but to fit well into the ecosystem.

46. Hereafter I refer to Marcus Aurelius Antoninus by his praenomen alone because it is the only name he bore throughout his entire life.

47. Marcus Aurelius, *Meditations*, trans. G. Hays (New York: Modern Library, 2003), *Med.* xi.8, 149–150. All quotations from Marcus are from this source.

48. Marcus, *Meditations* iii.2, 27–28.

49. Marcus, *Meditations* iv.48, 48.

50. Marcus, *Meditations* xi.35, 157.

51. Marcus, *Meditations* vii.59, 95.

52. Marcus, *Meditations* viii.51, 112.

53. Marcus, *Meditations* ii.17, iv.43, v.23, vi.15. See William O. Stephens, *Marcus Aurelius: A Guide for the Perplexed* (London: Continuum, 2012), 101–109.

54. Marcus, *Meditations* viii.15, 104; translation modified, emphasis added.

55. Epictetus, *Discourses* 4.8.36–40, my translation. Gardens of Adonis were domestic gardens dedicated to Adonis akin to greenhouses. To accelerate their growth, seeds were planted in porous clay pots or sponges where they rapidly and lushly sprouted but then also rapidly withered.

56. Epictetus, *Disc.* 1.6.12–22.

57. Marcus Terentius Varro, *De Re Rustica* i.2.6, quoted in K. D. White, *Roman Farming* (Ithaca, NY: Cornell University Press, 1970), 22.

58. See William O. Stephens, "Epictetus on Beastly Vices and Animal Virtues," in *Epictetus: His Continuing Influence and Contemporary Relevance*, ed. D. R. Gordon and D. B. Suits (Rochester, NY: RIT Press, 2014), 207–239.

59. Following the example of the ancient Greek poet Hesiod's didactic poem *Works and Days* c.700 BCE.

60. Cicero, *De Senectute, De Amicitia, De Divinatione*, trans. W. A. Falconer (London: William Heinemann, 1946), 4, 5, 6, 52, 56, 57–58, 63–65.

61. White, *Roman Farming*, 38–39; emphasis added.

62. Cicero, *On Duties*, trans. M. T. Griffin and E. M. Atkins (Cambridge: Cambridge University Press, 1991), *De Off.* 1.151, 58–59.

63. Cicero, *Pro Roscio Amerino*, 75.

64. Kelly Richman-Abdou, "How a 'Very Paintable' House Inspired 'American Gothic,' a Modernist Masterpiece," *My Modern Met*, September 29, 2019, https://mymodernmet.com/grant-wood-american-gothic/.

65. Richman-Abdou, "How a 'Very Paintable' House Inspired."

66. Richman-Abdou, "How a 'Very Paintable' House Inspired," using "stoic" here to mean small "s" stoic.

67. Note that Wood chose McKeeby to model Iowan agrarianism even though he was a dentist, not a farmer. This invites the possibility that Midwest (Stoic) agrarianism is a mindset, a Weltanschauung, a way of being and living in nature that need not exclude those who do not farm.

68. Wendell Berry, "The Agrarian Standard," in *The Essential Agrarian Reader*, ed. Norman Wirzba (Berkeley, CA: Counterpoint, 2003), 28, quoting Virgil, *The Georgics*, trans. L. P. Wilkinson (Penguin Books, 1982), 128.

69. Berry, "The Agrarian Standard," 29.

70. Berry, "The Agrarian Standard," 29–30.

71. And most of them are in Kansas, a Fringe Four state.

72. Similarly, an office worker or brothel owner who practices agrarian virtues by, for example, gardening on weekends and being a responsible shopper and consumer of food, would be ipso facto superior to a farmer who lacks agrarian virtues.

73. Slade and Lee, *The Midwest*, xvii.

74. Slade and Lee, *The Midwest*, 113.

75. Slade and Lee, *The Midwest*, 82.

76. Jonathan Foley, "It's Time to Rethink America's Corn System," *Scientific American*, March 5, 2013.

77. Slade and Lee, *The Midwest*, xxviii–xxix.

78. Slade and Lee, *The Midwest*, xxix.

79. Ted Genoways, "Gagged by Big Ag," *Mother Jones* (July/August 2013), https://www.motherjones.com/environment/2013/06/ag-gag-laws-mowmar-farms/.

80. Rebecca Hillel, Olivia Kraus, and Sally Spencer-Thomas, "Stoicism, Stress and Suicide among Farmers," *Dr. Sally Spencer-Thomas*, October 4, 2017, https://www.sallyspencerthomas.com/dr-sally-speaks-blog/2017/10/3/stoicism-stress-and-suicide-among-farmers. According to the ancient Stoics, suicide under specific, extreme circumstances could be justified for the sage, but never for the rest of us. A Stoic sage is free of all disturbing passions and immune to stress, including stress suffered by today's Midwest farmers. Midwest Stoicism offers a viable therapy for farmers wrestling with suicidal thoughts.

81. Chuck Jones, "Amid Trump Tariffs Farm Bankruptcies and Suicides Rise," *Forbes*, August 30, 2019.

82. Jeremy Rifkin, *Beyond Beef: The Rise and Fall of the Cattle Culture* (New York: Plume, 1993).

83. Slade and Lee, *The Midwest*, 114.

84. Jonathan S. Foer, *Eating Animals* (New York: Little, Brown and Co., 2009).

85. Michael Dent, "COVID-19: The Latest Zoonotic Disease Stemming from the Meat Industry," *IDTechEx*, May 7, 2020, https://www.idtechex.com/en/research-article/covid-19-the-latest-zoonotic-disease-stemming-from-the-meat-industry/20612.

86. The U.S. Global Change Research Program, *Impacts, Risks, and Adaptation in the United States: Fourth National Climate Assessment*, vol. 2, chap. 21, 2018, https://nca2018.globalchange.gov/chapter/21/. See also Union of Concerned Scientists, "Global Warming in the Midwest," updated June 19, 2019, https://www.ucsusa.org/resources/global-warming-midwest.

87. Charlene Arsenault, "How Climate Change Is Increasing Your Insurance Rates," Insure.com, September 4, 2019, https://www.insure.com/home-insurance/climate-change-home-insurance-rates.

88. USGCRP, *Impacts, Risks*, https://nca2018.globalchange.gov/chapter/21/.

89. Cato in Cicero's *De Senectute* and Marcus Aurelius in the *Meditations* are two examples. For the latter, see Stephens, *Marcus Aurelius*, 95–97.

90. Kenneth F. Kitchell Jr., *Animals in the Ancient World from A to Z* (New York: Routledge, 2014), 16.

91. Slade and Lee, *The Midwest*, xviii.

92. For examples of accounts of environmental virtue ethics, see Louke van Wensveen, *Dirty Virtues: The Emergence of Ecological Virtue Ethics* (Amherst, NY: Prometheus Books, 2000); Ronald Sandler and Philip Cafaro, eds., *Environmental Virtue Ethics* (Lanham, MD: Rowman & Littlefield, 2005); Robert Hull, "All about Eve: A Report on Environmental Virtue Ethics Today," *Ethics & the Environment* 10 (Spring 2005): 89–110; Ronald Sandler, *Character and Environment: A Virtue-Oriented Approach to Environmental Ethics* (New York: Columbia University Press, 2007); Brian Treanor, *Emplotting Virtue: A Narrative Approach to Environmental Virtue Ethics* (Albany, NY: SUNY Press, 2014).

93. Many peoples indigenous to what is now called the Midwest did not engage in the sorts of farming practices praised by the ancient Stoics. These Amerinds' activities involved hunting, gathering, and land management practices like prairie burns. But history cannot be reversed. Given today's agricultural realities, perhaps Midwest Stoic agrarianism is inextricably linked to the settler colonial ideal of the past. Going forward, though, new sustainable practices and smarter uses of technology are needed and so desirable to Stoic agrarians.

94. See William Petrocelli, "Voters in Wyoming Have 3.6 Times the Voting Power That I Have. It's Time to End the Electoral College," *Huffpost*, updated November 11, 2017,

https://www.huffpost.com/entry/its-time-to-end-the-electoral-college_b_12891764. Petrocelli writes: "To put it another way, the three electors in Wyoming represent an average of 187,923 residents each. The 55 electors in California represent an average of 677,355 each, and that's a disparity of 3.6 to 1."

95. See Sarah Conly, *One Child: Do We Have a Right to More?* (New York: Oxford University Press, 2015).

96. That is, anthropogenic climate change. See Dale Jamieson, *Reason in a Dark Time: Why the Struggle to Stop Climate Change Failed—and What It Means for Our Future* (New York: Oxford University Press, 2014).

97. Jonathan Haidt, *The Righteous Mind: Why Good People Are Divided by Politics and Religion* (New York: Vintage Books, 2013), 224–284.

SUGGESTED READINGS

Aurelius, Marcus. *Meditations.* Translated by Gregory Hays. New York: Modern Library, 2003.

Berry, Wendell. *The World-Ending Fire: The Essential Wendell Berry.* Berkeley, CA: Counterpoint, 2018.

Bookchin, Murray. *The Ecology of Freedom: The Emergence and Dissolution of Hierarchy.* New York: Black Rose Books, 1991.

———. *The Philosophy of Social Ecology: Essays in Dialectical Naturalism.* Montreal: Black Rose Books, 1990.

———. *Remaking Society.* New York: Black Rose Books, 1990.

Conly, Sarah. *One Child: Do We Have a Right to More?* New York: Oxford University Press, 2016.

Dombrowski, Daniel A. *The Philosophy of Vegetarianism.* Amherst: University of Massachusetts Press, 1984.

Epictetus. *Discourses, Fragments, Handbook.* Translated by Robin Hard. Oxford: Oxford University Press, 2014.

Foer, Jonathan Safran. *Eating Animals.* New York: Little, Brown and Company, 2009.

Jamieson, Dale. *Reason in a Dark Time: Why the Struggle to Stop Climate Change Failed—and What It Means for Our Future.* New York: Oxford University Press, 2014.

Johnson, Brian E. *The Role Ethics of Epictetus: Stoicism in Ordinary Life.* Lanham, MD: Lexington Books, 2014.

Kemmerer, Lisa. *Eating Earth: Environmental Ethics and Dietary Choice.* Oxford: Oxford University Press, 2015.

Long, A. A. *Epictetus: A Stoic and Socratic Guide to Life*. Oxford: Oxford University Press, 2002.

Lovejoy, Arthur O. *The Great Chain of Being: A Study of the History of an Idea*. Cambridge, MA: Harvard University Press, 1982.

Porphyry. *On Abstinence from Killing Animals*. Translated by Gillian Clark. London: Bloomsbury Academic, 2014.

Rachels, James. *Created from Animals: The Moral Implications of Darwinism*. Oxford: Oxford University Press, 1990.

Rifkin, Jeremy. *Beyond Beef: The Rise and Fall of the Cattle Culture*. New York: Plume, 1993.

Rufus, Gaius Musonius. *That One Should Disdain Hardships: The Teachings of a Roman Stoic*. Translated by Cora E. Lutz. New Haven, CT: Yale University Press, 2020.

Sandler, Ronald. *Character and Environment: A Virtue-Oriented Approach to Environmental Ethics*. New York: Columbia University Press, 2007.

Sandler, Ronald, and Philip Cafaro, eds. *Environmental Virtue Ethics*. Lanham, MD: Rowman & Littlefield, 2005.

Seneca, Lucius Annaeus. *Letters on Ethics*. Translated by Margaret Graver and A. A. Long. Chicago: University of Chicago Press, 2015.

———. *Natural Questions*. Translated by Harry M. Hine. Chicago: University of Chicago Press, 2010.

Simon, D. R. *Meatonomics*. San Francisco: Conari Press, 2013.

Stephens, William O. "Epictetus on Beastly Vices and Animal Virtues." In *Epictetus: His Continuing Influence and Contemporary Relevance*, edited by Dane R. Gordon and David B. Suits, 207–239. Rochester, NY: RIT Press, 2014.

———. "Five Arguments for Vegetarianism." *Philosophy in the Contemporary World* 1, no. 4 (1994): 25–39.

———. *Marcus Aurelius: A Guide for the Perplexed*. London: Continuum, 2012.

———. "Public Health, Ethical Vegetarianism, and the Harms of the Animal Food Industry." *Archives in Biomedical Engineering & Biotechnology* 2, no. 1 (2019).

———. "Stoic Naturalism, Rationalism, and Ecology." *Environmental Ethics* 16, no. 3 (1994): 275–286.

———. "Stoicism and Food." In the *Encyclopedia of Food and Agricultural Ethics*, edited by David M. Kaplan. Dordrecht: Springer, 2018.

———. "The Stoics and Their Philosophical System." In *The Routledge Handbook of Hellenistic Philosophy*, edited by Kelly Arenson, 25–37. New York: Routledge, 2020.

Stuart, Tristram. *Waste: Uncovering the Global Food Scandal*. New York: W.W. Norton & Co., 2009.

Treanor, Brian. *Emplotting Virtue: A Narrative Approach to Environmental Virtue Ethics*. Albany:

State University of New York Press, 2014.

Usher, M. D. *Plato's Pigs and Other Ruminations: Ancient Guides to Living with Nature*. Cambridge: Cambridge University Press, 2020.

Walters, Kerry S., and Lisa Portmess, eds. *Ethical Vegetarianism from Pythagoras to Peter Singer*. Albany: State University of New York Press, 1999.

Wensveen, Louke van. *Dirty Virtues: The Emergence of Ecological Virtue Ethics*. Amherst, NY: Prometheus Books, 2000.

Whiting, Kai, William O. Stephens et al. "How Might a Stoic Eat in Accordance with Nature and 'Environmental Facts'?" *Journal of Agricultural and Environmental Ethics* 33, no. 3 (2020). .

Williams, Gareth D. *The Cosmic Viewpoint: A Study of Seneca's* Natural Questions. New York: Oxford University Press, 2012.

Wirzba, Norman, ed. *The Essential Agrarian Reader*. Berkeley, CA: Counterpoint, 2003.

Zatta, Claudia. *Interconnectedness: The Living World of the Early Greek Philosophers*. Sankt Augustin, Germany: Academia Verlag, 2017.

Searching for the Just Shrinking City in Flint, Michigan

Benjamin J. Pauli and Levi Tenen

Midwestern Cities and Justice in the Era of Urban Crisis

Midwestern cities have long signified both the promise and the perils of the urban environment. Their stunning growth from "small villages, barely more than fortified outposts,"[1] speckling the settler frontier to industrial powerhouses and oases of middle-class affluence made them "the wunderkinds of the American family of cities" in the late nineteenth and early twentieth centuries.[2] Civic leaders sought to make these cities, in David Mallach's words, "true cities, not merely in the sense of large agglomerations of people and business activity but, following as best they could the models of ancient Athens or renaissance Florence . . . centers of civic, cultural, and intellectual life."[3] As developers and city planners funneled the fruits of economic growth into cultural institutions, beautification initiatives, and modern infrastructure, the average Midwestern city-dweller came to enjoy a standard of living equal to or greater than anyone else in the world.

By the middle of the twentieth century, however, the fabric of Midwestern urban life had begun to unravel. The rise of the mass-market automobile and the construction of car-friendly roads shifted residential and commercial development to the suburbs, as more and more people opted to commute to work from outside city limits. During the civil rights era, the eruption of latent racial tensions spurred an exodus of white residents from urban centers, while the allure of higher profit margins in places like the American Southwest, Mexico, and China led many industries to shutter factories and relocate, driving up unemployment, poverty, and crime. By the 1970s, the triumphalist narrative that held up Midwestern cities as the avant-garde of American civilization had been flipped on its head: now cities like Detroit, Michigan; Cleveland, Ohio; and Gary, Indiana, were the poster children of decline, their troubles symptomatic of a broader "urban crisis."[4]

While these cities have not "died" outright, as has sometimes been prophesied, they continue to struggle with profound structural challenges. This is especially true of so-called shrinking cities, where population levels are falling and, in some cases, stand at less than 50 percent of their post–World War II apexes.[5] Population loss can result in dramatic changes to the character and financial foundation of the urban environment. Shrinking cities of the Midwest tend to have high rates of vacancy and blight, which correlate with other problems like arson and crime and drive down property values, eroding a city's tax base. The loss of revenue from income and property taxes can make it difficult for a city to offer basic services like police and waste disposal, provide pensions and health care to retired city employees, and maintain (much less upgrade) aging infrastructure. To balance their budgets, shrinking cities have laid off personnel, outsourced services, privatized public assets, and invited controversial forms of development in the hopes of generating new sources of revenue.

Still, many urban-studies scholars argue that shrinking cities have yet to come fully to terms with the stark realities confronting them. Policies adopted in response to disinvestment and population loss often seek to stave off further decline or even restore former patterns of growth, as if the sweeping economic, political, and cultural shifts at the root of the urban crisis can be counteracted through savvy municipal-level decision-making. Instead, many scholars have advocated replacing the growth-oriented idealism of the past with what has been called a "new realism."[6] While such realism comes in different forms, the general idea is that shrinking cities should accept that decline is inevitable for the foreseeable future and concentrate on declining "smartly."[7] If cities were to put more thought into *how* to decline, so the thinking goes, they could mitigate its downsides and possibly even find opportunity in crisis, actually enhancing residents' quality of life in certain respects. As Justin Hollander writes,

> Population loss does not have to be a bad thing for a community . . . [L]ess people can mean less congestion and leading [*sic*] to more open space and recreational opportunities. Less people in a city can mean lower student to teacher ratios in the classroom and faster response times from the police and fire departments. A smaller population has a smaller environmental footprint, helping a city to reduce its impacts on global climate change. Fewer jobs mean fewer cars on the roads and improved air quality. Fewer factories mean less pollution and better water quality. While population and employment declines are never the aim of public policy and planning (and are usually associated with negative outcomes), such declines, if managed properly, can be potentially beneficial to residents and workers left behind.[8]

If this is realism it is, at least, hopeful realism, offering a vision of cities that have shed some of the most objectionable elements of their industrial pasts. Indeed, many of the benefits mentioned above seem to flow almost automatically from decline, simply by virtue of subtracting people and pollution. Hollander's caveat about proper "management," however, reflects the idea that local governments must engage in some conscious steering of the process if it is to prove advantageous on the whole.

It is the argument of this chapter that if shrinking is to result in a city that residents might actually see as an improvement in certain respects, then it is critical that managers, planners, and city residents themselves strive to develop an idea of what we call the *just* shrinking city. This is particularly important because of the long history of *in*justice in these communities, and the real possibility that any embrace of shrinking will be seen as yet another form of injustice. A just shrinking city, as we define it, is one in which shrinking is in some sense accepted, but in a way that foregrounds considerations of justice and seeks opportunities to advance justice when possible through certain kinds of managed shrinking.

It is unlikely that a fully articulated idea of this kind will arise solely from the deliberative, participatory planning processes recommended by contemporary planners, at least as those processes are currently imagined. However, this fact makes the need for conceptions of a just shrinking city no less significant; rather, it simply means we must carefully consider how they can be developed without disrespecting the agency and interests of shrinking city residents. We will explore the relevance of our argument to a specific shrinking-city context: pipe-replacement efforts in Flint, Michigan, following that city's notorious water crisis, which have given rise to a debate about whether an opportunity was missed to "rightsize" water infrastructure in the wake of the crisis. While both sides of that debate pick up on important considerations, they ultimately reach a stalemate, we argue, because not enough attention has gone into exploring whether and how a rightsized Flint could be a just outcome. The concept of the just shrinking city can, in our view, not only help to break that stalemate, but perhaps also offer a constructive way of imagining a future for Midwestern cities beyond the urban crisis.

The Insufficiency of Procedural Justice in the Shrinking Urban Environment

Whenever we consider the proper "management" of decline, we are confronted by the question *who*, exactly, is managing the decline? Who decides where opportunity lies, or when a dose of "realism" is in order in a shrinking-city context? Typically, responsibility for management at the local level lies with city administrators appointed by mayors or hired by city councils and professional planners within city planning departments. In more extreme cases of financial insolvency, individual "emergency" managers may be temporarily empowered by bankruptcy judges or state governors to take control of city affairs. Ever since the failure of top-down urban renewal initiatives and the reinvigoration of ideals of participatory democracy in the 1960s and 1970s, however, monopolization of city management and planning by officials and professionals has come under robust critique.[9] It is now widely accepted that those who reside in a city (shrinking or otherwise) have a right to be involved at some level in planning and policy decisions affecting their well-being. Furthermore, even the most capable and informed managers and planners are unlikely to possess all the ability and information they need to create effective plans by themselves. This is, perhaps, especially true in shrinking cities beset by so-called wicked problems—problems that are complex, open-ended, and interconnected, and that do not lend themselves to straightforward technocratic solutions. Under such conditions, the knowledge and ideas that ordinary residents have to offer can be critical in charting a viable path forward.

In light of the normative and practical limitations of management from above, even those who advocate a smart shrinkage approach to urban decline do not necessarily trust themselves to formulate ideal end states at which smart shrinkage should aim. Instead, Hollander and colleagues argue that planners should focus on developing just procedures by which planning options are discussed and chosen among resident populations, incorporating principles like inclusion, deliberation, recognition, transparency, and scale appropriateness (ensuring that plans are "local in control and implementation" even when they are "regional in scope").[10] In theory, plans created using just procedures will themselves be seen as just, and will thus be more acceptable to those impacted by them.[11] At its extreme, a procedure-oriented approach conceives of just ends in a purely formal way: that is to say, they are what emerge through an inclusive, transparent (and so on) planning process, with their content left open to whatever a deliberative group decides.

While we agree that the procedural norms mentioned above are important, we do not think they eliminate the need for considering what just outcomes should look like in shrinking cities independently of the process by which those outcomes are reached. To begin with the obvious, many states of affairs within shrinking cities come about by means other than just procedures. Changes to the urban environment might arise, for example, through the operation of market forces, or through changes to the natural environment such as global warming, or simply through decisions made by city administrators without public involvement, or through deliberation that not everyone recognizes as being legitimate. Unless we are willing to say flatly that non-participatory, non-deliberative outcomes are by definition unjust, we will need to have some way of thinking about what a just shrinking city ought to look like independently of procedural considerations.[12]

Secondly, even within participatory, deliberative contexts, it is inevitable—and perhaps even desirable—that prior ideas about justice are brought in from outside. It is likely that most if not all participants in a deliberative process will already hold opinions about what justice entails, and these ideas may end up being more influential within the process than those developed during deliberation itself. Indeed, those who exercise disproportionate influence over deliberative processes—those who convene them, structure them, facilitate them, or exercise de jure or de facto leadership within them—sometimes have a fairly developed sense of what a just outcome would look like, and may attempt to steer the process in various ways toward a particular outcome. Although excessive steering would clearly violate norms of procedural justice, it is not necessarily counterproductive for those with more refined visions of just ends to play a larger role than others in designing deliberative processes and shaping the conversations that emerge within them.[13] Rather than insisting that ends be devised in a strictly egalitarian fashion, we may insist, simply, that part of the "transparency" of a deliberative process is for those with preexisting views to bring them out into the open rather than engaging in tacit manipulation of the process—especially people who are in a position to exert more sway than others.

To be sure, any concessions we make to inequality within deliberation must be made with great caution and careful justification. Just because some views are more developed does not necessarily mean that they are complete and that those who hold them cannot learn from the collective knowledge marshaled within deliberative bodies. The more refined ideas of, say, a professional planner are also not inherently better than views that draw from intuition and personal experience simply by virtue of being refined. Space must be made within deliberation for all kinds of views, proposals, and ways of expressing them. But this does not mean that we can—or should—deny entirely the special role of those with more fleshed-out ideas of desirable end states. We should be thoughtful, instead, about how and when we allow or encourage them to bring their views into deliberative settings, or to act on them in other contexts.

A further consideration is that subordinating ends to means, as proposed by proceduralist frameworks, may be especially unlikely to produce viable outcomes in shrinking-city contexts—that is, by our measure, outcomes that are both "realistic" and "just." In a well-resourced community, it may be possible to develop plans in a participatory manner that both embody widely held notions of justice and are "realistic"—financially speaking and otherwise. In shrinking-city contexts, however, there may be a large gap between what a professional planner argues is necessary to confront reality and what residents think is right or just. Planners, as we have already indicated, are likely to view continued decline as inevitable in these cities and as necessitating radical breaks with past practices, as well as dramatic reconfiguration of urban spaces and institutions. Their proposals for shrinking cities often call for some measure of "rightsizing": the paring back of city services and aspects of the built environment such as roads, pipes, and parks that require management and maintenance.[14] Residents of shrinking cities, by contrast, tend to harbor less fatalistic views and a more ambitious sense of possibilities for underused and dilapidated urban spaces. Their attitudes often reflect a strong sense of place attachment that makes them loath to give up on, or fundamentally repurpose, such spaces. They may, in fact, envision the restoration of cherished features of the city that have been lost and the revival of abandoned residential and commercial areas. Similarly, they may see cutbacks to city services and reorganization of city institutions as threats to the basic identity and purpose of a city—as, in effect, killing a city in order to "save" it.

Especially in cities where population loss can be traced back to dynamics of structural racism and economic exploitation, it may, from a resident perspective, reek of long histories of injustice, and accepting or embracing shrinkage in any way may be seen as making oneself complicit in this injustice. Furthermore, it is well known that policies aiming to improve the financial standing of government through cost-cutting measures have often been the source of new injustices. Residents may understandably see rightsizing as an expression of what has been called the neoliberal turn in urban governance, involving the dismantling of public institutions and public spaces and the concomitant rise of development corporations and other private interests intent upon remaking cities in their own image. Insofar as private interests are positioned to take advantage of the opportunities that are created by population loss, the result is likely to be cities designed for the tastes and lifestyles of those with the most purchasing power. Poor residents and residents of color in particular risk becoming even more marginalized economically, politically, and geographically, leading to what Mallach has called the "divided city," a place of "growing inequality, increasingly polarized between rich and poor, white and black."[15] Additionally, insofar as government plans to rightsize involve the teardown of existing structures and the relocation of residents, they readily bring to mind the bad old days of urban renewal.

Thus, residents of shrinking cities tend to be skeptical that intentional shrinkage (à la rightsizing) to complement population loss will produce just outcomes. What this would seem to mean is that the more just the planning process—that is, the more sensitive it is to popular influence and control—the less likely it will be to result in plans that embrace shrinking. On the surface, this logic may appear to be contradicted by the development in recent years, through participatory procedures, of shrinking-city master plans that involve a certain amount of rightsizing, as in Youngstown (*Youngstown 2010*), Detroit (*Detroit Future City*), and Flint (*Imagine Flint*). However, while these plans make certain concessions to the changing urban landscape, their rightsizing proposals tend to be modest in scope (and notably to avoid the language of "shrinkage" altogether).[16] In *Imagine Flint*, for example—the drafting of which involved input from around five thousand residents—outright decommissioning is recommended only for water infrastructure under abandoned industrial sites, one isolated (and unused) park, and a few dilapidated dams on the Flint River. Even if these kinds of measures result in some cost savings, they are unlikely to move the needle very far in extricating Flint or any other shrinking city from the financial dilemmas they face. In other words, shrinking-city master plans typically do not propose to scale back the built environment to anything approaching what an advocate of realism might deem to be the right size.

Additionally, if the fate of the *Youngstown 2010* plan is any indication, popular resistance to even modest proposals for deliberate shrinkage may not be overcome simply because these proposals are incorporated into plans devised through relatively participatory processes. In their analysis of zoning in Youngstown subsequent to adoption of the city's master plan, Ryan and Gao found that locals remained attached to "growth-oriented culture," individual property rights (as trumping collective plans), and "social values" discouraging radical changes to traditional land uses—all of which became obstacles to implementing *Youngstown 2010*.[17] What these persistent attitudes seem to suggest is that Youngstown residents have not been fully sold on the vision articulated in their master plan. Given indications that the same is true of residents in other shrinking cities, it appears that we are still in need of a vision of a *just* shrinking city robust enough to win over both planners and the public.

Toward the Just Shrinking City

In order for managers, planners, and residents alike to develop a conception of the just shrinking city, it is of course paramount to have a sense of what justice is. However, "justice" has become a notoriously vague and capacious concept, connoting a wide range of values and priorities depending on who is invoking it. Philosophical efforts to formulate ideal theories of justice have attempted to bring some coherence and definition to the concept, but often hover at such an extreme remove from on-the-ground issues of everyday life that they are of limited practical application. As Charles Mills has argued, theories of justice are typically based on idealized and ahistorical conceptions of social institutions and of people's capacity for assessing political principles, offering few resources for navigating *im*perfect realities.[18] Clearly, efforts by philosophers to imagine ideally just cities (as in Plato's *Republic* and its many imitators) are vulnerable to the same criticism, despite the fact that they do, in some sense, grapple with the relationship between justice and questions of urban form.

Contemporary environmental ethicists have made more concrete attempts to theorize the "built" environment, including the urban landscape, and to explore its relationship to normative principles.[19] Some of these efforts seek to determine what kind of city constitutes a "natural" environment for human beings.[20] Ingrid Leman Stefanovic, for example, writes that "A natural city is one that respects diversity; one that arises organically; one that invites local community engagement; one that respects not only ecological limits but the richness and diversity of historically grounded, ontological roots of human well-being."[21] The broader point is that the decisions human beings make about how to arrange urban space and institutions "[reflect] who we are" and embed normative "human judgments" in the built environment. As Roger King writes:

> Construction of the built environment—design of buildings and cityscapes, creation of artifacts, organization of the agricultural landscape—shapes materials, arranges them in patterns, establishes new relationships among them and between them and their users. In the process, human beings take a stand in the world; their contrivances amount to judgments of what is appropriate, significant, trivial, or irrelevant.[22]

Rarely have so-called urban environmental ethicists considered the arrangement of urban environments from the standpoint of justice specifically, however—at least not justice in a positive sense. What *has* become common is to speak of its opposite, environmental *in*justice, with reference to the city. Environmental injustice typically refers to disparate impacts on poor people and people of color caused by contamination of "natural" substances that undergird and flow through the city (like soil, air, and water) and by deterioration of the built environment (like buildings and infrastructure). Thanks in no small part to the Flint water crisis, attention to matters of environmental injustice has exploded in recent years, and the catalog of all the various ways in which people have been harmed by degradation of urban environments has expanded considerably.

However, an environmental justice framework offers at least some guidance for what a just environment would look like. At a minimum, a just environment would be characterized by a fair distribution of environmental "goods" and "bads":

if there are to be polluting facilities in a city, for example, their impacts should at least be spread around equitably rather than concentrated in areas where people are already worse off. (Conversely, access to breathable air should not be limited to those in more privileged neighborhoods.) Additionally, calls for environmental justice often go beyond merely arguing for fairer distributions by implying that certain kinds of environmental bads should be eliminated entirely. It may be that the most environmentally just city is one without *any* meaningful pollution of water, soil, and air, without dilapidated houses and roads, and so on.

Envisioning (environmental) justice within the city as the removal or inversion of environmental injustice is not wholly adequate, however. When conceptualized positively, justice may have elements that do not correspond in any direct way to existing injustices. Rather, it may involve enhancing or introducing any number of features to the city that improve the overall quality of life for residents.[23] Attempting to imagine all the various enhancements that might fall under this umbrella, however, can easily become overwhelming. For example, the Just City Lab, part of the Harvard Graduate School of Design, has compiled a list of no fewer than forty-nine "values indicators" organized into twelve categories to capture what justice looks like within the design of urban environments:

Acceptance: Belonging, Empathy, Inclusion, Reconciliation, Respect, Tolerance, Trust

Aspiration: Creative innovation, Delight, Happiness, Hope, Inspiration

Choice: Diversity, Spontaneity

Democracy: Conflict, Debate, Protest

Engagement: Community, Cooperation, Participation, Togetherness

Fairness: Equality, Equity, Merit, Transparency

Identity: Authenticity, Beauty, Character, Pride, Spirituality, Vitality

Mobility: Access, Connectivity

Power: Accountability, Agency, Empowerment, Representation

Resilience: Adaptability, Durability, Sustainability

Rights: Freedom, Knowledge, Ownership

Welfare: Healthiness, Prosperity, Protection, Safety, Security

This enumeration of values is supposed to facilitate "value-based aspiration for urban stabilization, revitalization and transformation" and encourage textured, imaginative thinking about the many elements of the "just city."[24] But while it seems to reflect a noble effort to account for the complex dynamics of urban environments and be inclusive of a wide range of priorities, it also risks stretching the concept of justice so thin that the latter becomes almost all-encompassing. If creative innovation, protest, and spirituality are all components of the just city, one might wonder what would justify keeping any particular desideratum *off* the list—indeed, it is possible, and perhaps even tempting, to expand such a list further in order to be even more inclusive. One might also ask how one is supposed to balance such a large number of considerations in real-world contexts where values will inevitably clash at times. How do we know which value to prioritize if conflict begins to undermine togetherness, or empathy, or happiness, or any number of other values? How do we reconcile spontaneity and adaptability with durability, or equality and equity with ownership? Which of the forty-nine values should a planner refer to when making any particular design decision?[25]

More targeted approaches that focus on especially important values arguably make justice more actionable. The influential account of the just city developed over the past fifteen years by Susan Fainstein offers a useful contrast to the lab's approach in this respect (though her account is part of the inspiration for the Just City Lab indicators). Drawing from philosophical theories of justice, Fainstein limits her delineation of urban justice to a far more manageable list of three core values: democracy, diversity, and equity. She then derives from these values fairly specific prescriptions. For instance, she argues that cities seeking to realize the value of diversity should ensure that land use is mixed and that there is "ample," "widely accessible," and "varied public space." Cities concerned with equity should, among other things, insist that all new housing developments include affordable units and that new commercial development "facilitate the livelihood of independent and cooperatively owned businesses." However, Fainstein does not directly consider whether and how her approach applies to *shrinking* cities in particular (notably, in *The Just City* her case studies are the *growing* cities of New York, London, and Amsterdam).[26]

While not attempting anything like a full elaboration of the just shrinking city, Alan Mallach is one of the few to gesture in that direction, offering two examples of what shrinking cities should strive for as a just end state. He argues, firstly, that "a just economic development strategy" aiming to rejuvenate economic life in declining cities "must be to some extent redistributional." For instance, a proposal to bring in new jobs that caters only to "highly educated immigrants to the city" and fails to provide opportunity to existing residents would be unjust. However desperate shrinking cities may be for investment, they should ensure that economic goods like job opportunities are available to everyone, rather than counting on the benefits of development to trickle down to economically marginalized residents indirectly. Mallach also argues that cities should avoid rightsizing if doing so would impair access to resources and opportunities for those residents who choose to remain in their current neighborhood. A city's rightsizing policies should not, in other words, make residents remaining in depopulated parts of the city any worse off than they already are.[27]

Mallach's suggestions are a start, but clearly there is much work to do in fleshing out what justice looks like under the especially challenging circumstances faced by shrinking cities. While proposals for enacting justice typically involve the subtraction of bads (like pollution and discrimination) and the addition of goods (like diversity and job opportunities), shrinking cities sometimes have little choice but to subtract *goods*, eliminating or repurposing aspects of the city that residents still believe to be good or capable of becoming good again. Shrinking cities may also feel compelled to adopt policies that disrupt patterns of life in unwelcome ways—for example, by encouraging relocation despite residents' feelings of place attachment. Furthermore, policies that prioritize some neighborhoods over others may contradict proud local histories and undermine residents' sense of community, identity, and purpose. It is difficult, perhaps, to see how such measures can be fully consistent with principles of justice, much less how they might advance the implementation of those principles in any way. We might rightly wonder, then: is it even possible for shrinkage to result in a just outcome?

While we cannot answer this question definitively in this chapter, nor offer a fully articulated portrait of a just shrinking city, we hope to demonstrate why it is an important question to ask, and how it can be employed productively within a concrete shrinking-city context. In so doing, we illustrate some of the considerations and conundrums that arise in the pursuit of justice under shrinking-city conditions.

Pipe Replacement in Flint

In April 2014, an emergency manager appointed by Michigan governor Rick Snyder to shepherd the city of Flint out of a stubborn financial crisis oversaw the temporary switch of Flint's water source from Lake Huron to Flint River water. Famously—or better, infamously—the change wreaked havoc on the city's water system. Failure to treat the water properly at Flint's understaffed and underequipped water treatment plant led to a variety of problems with water quality and, ultimately, public health: rashes, hair loss, a historic outbreak of Legionnaires' disease, and lead poisoning, as the water ate away at the protective scale on lead pipes ("service lines") connecting homes and other buildings to water mains. The nation watched in horror—that is, after months of resident activism exposed what was happening—and the crisis was denounced far and wide as an egregious, indeed, historic example of environmental injustice.[28]

The Flint water crisis has been held up as a cautionary tale about the consequences of adopting a neoliberal approach to challenges faced by shrinking cities: while decades of accumulated structural vulnerabilities had set the stage for disaster, it was cost-averse, austerity-minded technocrats—emergency managers—who pushed Flint over the edge. Their prioritization of finances, their insistence on cutting or minimizing expenses even in the face of potential threats to the public, put Flint on a path toward using an inferior but cheaper water source and left it ill-prepared to confront the public health crisis that resulted.[29] The main lesson of the water crisis, then, from this perspective, is that it is necessary to find other ways out of the *urban* crisis—ways that involve bigger, broader, and more principled thinking than the financial myopia of number-crunchers determined to force cities into living within their means.

In fact, it was a mere six months prior to the switch in water source that the Flint city council adopted the aforementioned *Imagine Flint*, a thirty-year plan meant to embody exactly the kind of big thinking necessary to chart Flint's course out of its postindustrial predicament. Over a period of three years, thousands of residents had met and deliberated in meetings large and small about the future of Flint. The process leading up to the plan was widely celebrated (receiving the 2014 Planning Excellence Award for Public Outreach from the Michigan Association of Planning), suggesting that at least some measure of procedural justice had been realized. The plan itself also incorporated efforts to think about what "equitable" ends would look like in Flint, making explicit reference to promoting "social justice" by, for example, "developing public and affordable housing" and by ensuring accessibility of parks and green space to all residents.[30]

Despite the participatory process that produced *Imagine Flint*, the fair amount of goodwill behind it (the council adopted it unanimously), and its explicit effort to foreground considerations of equity and social justice, the plan included some elements that were not entirely uncontroversial, such as proposals to redesignate for zoning purposes the city's most depopulated neighborhoods in the hopes of encouraging their gradual transition to green space. These proposals came in the wake of prior suggestions by city leaders that the city consider "shutting down" entire "quadrants," in the words of interim mayor Mike Brown.[31] Arguing that a return to Flint's industrial glory days was "not gonna happen," former Genesee County Land Bank executive (and current congressman) Dan Kildee made it clear that drastic moves in the direction of shrinkage were necessary even if "not everyone" would "win."[32] While Brown's and Kildee's calls for radical rightsizing did not find their way into *Imagine Flint*, the coverage and controversy they inspired ensured that they continued to echo in the background of the master planning process, particularly when shrinkage was proposed in any form. Whatever consensus was achieved around *Imagine Flint*, then, there remained the risk that it would be shattered if residents came to see its proposals as vehicles for the more sweeping changes sought by some Flint elites.

The latent fears that some residents had about the "real" intentions of the master plan were brought out into the open during the water crisis. Some residents hypothesized that the crisis itself was orchestrated—with the help of the emergency manager system—to drive inconvenient people out of the city, or at the very least, that the crisis recovery effort would be used as an opportunity to push forward with radical shrinkage by, for example, writing off large parts of the damaged water system. In light of these fears, the call for justice coming from residents and water activists was to *leave no one, and no place, behind*. Everyone, they insisted, had a right to clean water straight out of the tap—with all the infrastructure and services that necessitated—regardless of who they were or where they happened to live in the city.

Reflecting this principle, one of the core demands of residents and activists was that water pipes be replaced everywhere, including every lead service line in Flint's water system. More than any other infrastructural issue brought to light by the water crisis, the issue of lead service lines (LSL) removal became the focus of concern in Flint and a leading priority within the crisis response beginning in 2016. Flint's 10,000-plus lead pipes were the legacy of its early-twentieth-century boom years, when the founding of the General Motors (GM) corporation on the north bank of the Flint River (c.1908) and the explosion of the auto industry drew tens of thousands of new residents to the city. As was typical of Midwestern cities, most of these residents ended up in freestanding, single-occupancy houses—in this case, courtesy of GM itself, as it financed the construction of new residential neighborhoods across the northern half of the city. The water system had to grow, of course, to accommodate this development. While iron-based water mains formed the system's spine, lead was the material deemed most suitable to connect houses and other buildings to the expanding grid: it was durable, cheap, and flexible, allowing pipes to be woven around tree roots and other idiosyncrasies of individual yards. But it was also toxic—as we now know, even at vanishingly low levels of exposure.

Lead pipes were, in effect, the poison straws through which Flint residents consumed their water for the better part of a century. Not that residents even knew of their existence, necessarily: among the revelations of the water crisis for many was the simple (and counterintuitive) fact that lead was used in plumbing at all. Some took pennies and magnets into their basements to test the piece of pipe coming into their water meters through the wall, finding, more often than not, that it was indeed made of the dreaded material. What's more, residents learned that their service line (the length of it from the sidewalk to the house, anyway) was *their* personal property rather than the city's, and typically considered the property owner's responsibility. Indeed, in other cities, such as Madison, Wisconsin, which replaced its LSLs over a period of eleven years beginning in 2001, residents were asked to cover the cost of replacing their side of the line. Fortunately for Flint residents, they were ultimately spared this cost by a windfall of state and federal crisis-response money, which, among other things, funded comprehensive replacement of LSLs. Thus, between 2016, when LSL excavations began, and 2021, when they ceased, Flint replaced its lead pipes faster than any American city had before.[33]

The speed and scope of the replacements, carried out under the glare of the national spotlight, offered inspiration to the many other cities across the country plagued by lead—above all, to other Midwestern cities, where the lion's share of LSLs are located.[34] Moreover, Flint illustrated the importance of extramunicipal support for LSL replacement: while cities were feeling more than ever the urgency of getting the lead out, they would not be able to do it alone. As excavations in Flint entered their final phase, the newly elected Biden administration announced that federal funding for comprehensive LSL replacement would be a cornerstone of its proposed infrastructure plan. This newfound sense of urgency around LSL removal, evident from the federal down to the local level, reflected the burgeoning consensus that traditional methods of mitigating lead-in-water exposure—such as corrosion control treatment, filtration, flushing, and so on—are inadequate. Ridding cities of their lead pipes entirely and preventing another Flint-style tragedy was now seen as a public health imperative—and a pressing matter of environmental justice, given lead's disparate impact on poor people and people of color. The response to Flint became a kind of annunciation that lead was no longer to be tolerated anywhere in the urban environment.

However, despite the fact that speedy, universal LSL replacement clearly reflected Flint residents' sense of justice in light of the water crisis, it had implications that went well beyond the immediate crisis response. Replacing pipes with new pipes while leaving intact the existing arrangement of a water system has the effect, of course, of further entrenching that system. Given the lifespan of an average water pipe (some one hundred years or more), such a process effectively locks a city into current configurations of urban space, on a timescale extending far beyond even the most ambitious master plan. In Flint's case, the replacement program helped to ensure that the city's pattern of residential settlement, characterized by single-occupancy houses spread across a large geographical area, would persist for some time into the future. This translates into an ongoing need for an extensive water grid, one increasingly riddled with dead zones as water use declines, difficult to manage and expensive to keep up, with more pipe per capita to maintain than in cities with higher density.

Thus, some have argued that the LSL replacement program should not have occurred without more careful thinking about the potential downsides of preserving a water system the size of Flint's. University of Michigan-Flint planning professor Victoria Morckel argues that Flint should *not* have "replicate[d] an oversized water system with piecemeal, one-for-one replacements" but instead should have used its water crisis as an opportunity to reconfigure its water infrastructure so as to reduce its capacity and spread.[35] In its effort to address immediate threats and satisfy popular demands, the LSL program effectively, in Morckel's view, made it harder to have a further-reaching discussion about whether the current configuration of the city *should* be preserved, shifting attention away from discussion of the long-term ends that Flint ought to move towards—the sort of discussion that was being had on the eve of the water crisis during the *Imagine Flint* master planning process.

Morckel maintains that using the water crisis as "an opportunity to right size" Flint's water infrastructure would have been consistent with the modest rightsizing proposals of *Imagine Flint*.[36] With the help of hydraulic modeling, she suggests, different arrangements of pipes, as well as smaller pipe sizes, could have been considered as a means of reducing maintenance costs, avoiding water stagnation, and encouraging (or discouraging) certain kinds of development above ground. However, parts of her argument seem to hint at more radical possibilities not as obviously consistent with either the (ostensible) aims of the master plan or with popular demands for justice in light of the crisis. While it is not clear that she thinks the city could have eliminated whole "quadrants" of its water system, or should have denied residents new service lines, she does imply that some extremities and pockets of the system might have been able to be pruned entirely.

It is perhaps for this reason that her invocation of rightsizing as an "opportunity" has generated considerable controversy. In a sharp critique, Michigan State University geographer Rick Sadler and coauthors accuse Morckel of advocating a discredited technocratic approach to planning and of ominously (though implicitly) raising the prospect of forced relocation of residents reluctant to abandon their neighborhoods. While their critique of Morckel reads into her argument more than appears to have been intended, it does raise legitimate questions about how any meaningful rightsizing proposal could be implemented in Flint without inflicting further injustice in both process and outcome. They point out, for example, that very few city blocks in Flint are sufficiently vacant to eliminate entirely the water infrastructure below; almost all have at least one occupied residence. Cutting off Flint's water infrastructure in almost any conceivable part of the city would, then, seem to require residents to move. Although, as Morckel points out, many Flint residents are already moving out of troubled neighborhoods of their own accord (indeed, between 2010 and 2020, the city lost 20 percent of its population), and more have expressed a desire to move, Sadler et al. suggest that the kind of rightsizing she hints at would go beyond anything made possible organically through population loss. Enough residents remain committed to even the worst-off parts of the city that implementing a shrinking agenda involving relocation would require overruling popular sentiment. And it would come at a substantial and unfair cost to those affected, "impos[ing] a greater burden on African American residents than on White residents."[37] Indeed, rightsizing in Flint, they claim, would constitute yet another environmental injustice.

Our contention is that both Morckel and Sadler et al. raise important issues, and that the two sides of this debate are not as far apart as they may appear at first glance. Morckel, in her response to Sadler et al.'s critique, acknowledges the importance of involving residents in any decisions made about rightsizing. Sadler et al., for their part, do not deny the significance of Flint's financial troubles or rule out the possibility that residents could, at some time in the future, come to see certain forms of rightsizing as sensible and just. What is needed, it seems, is a concept that strives to reconcile the "realism" and long-term thinking advocated by proponents of rightsizing with the need to respect popular self-determination and demands for justice. This is precisely what the concept of the just shrinking city offers. In proposing this concept, we do not mean to imply certainty that such a city can in fact be realized—rather, our intention is to steer controversies like the one described above around their current impasses and toward mutual exploration of the possibility of such a city.

What would it look like to consider water infrastructure from this perspective? To offer an illustration, let us return to Fainstein's conception of the just city. Recall that in her view, equity and diversity are essential components of just city outcomes. These values present an opportunity to evaluate the relative justice of infrastructure proposals in Flint—with sensitivity, of course, to the special considerations that arise in shrinking-city contexts.

Diversity

With respect to the relationship between water infrastructure and diversity, the question raised by a just shrinking city framework is the following: can a city maintain, much less promote, diversity while reducing or reconfiguring its water system to cope with population loss? There would seem to be some ways in which it can. Rightsizing and the reconfiguration of city infrastructure can, for example, allow residents to reimagine and re-create the urban environment in ways that increase the diversity of land uses. In contrast to conventional grid-like development, infrastructural reconfiguration underground, combined with creative zoning above ground, can help to open up new residential, commercial, industrial, and recreational possibilities—indeed, this is one of the objectives of the *Imagine Flint* plan, which calls for "equitable access to a diverse mix of uses."

However, whether new land uses benefit residents or private interests depends on how these possibilities are distributed. *Imagine Flint*, for example, created an entirely new zoning category called "Green Innovation," meant to help over-whelmingly depopulated neighborhoods transition into spaces that can be used for nonresidential purposes, such as agriculture, food processing, and renewable energy. While the plan frames Green Innovation areas as exciting opportunities for experimental uses of urban space, residents of those neighborhoods have expressed considerable fear and anxiety that they will be pressured in various ways to leave their homes, including through the neglect of infrastructure, to make way for new industries.[38] At the same time, some residents have found creative ways to make use of vacant land in Green Innovation zones, including creating their own green spaces and urban gardens.[39]

Shrinking may also offer unique opportunities to accommodate residents who would prefer to get their water off-grid, expanding diversity of choice in water provision. Usually it is assumed that justice in urban water distribution is best served by connecting every household to a universal public system, given that this is generally the most effective way to get the most residents consistent, safe, and (relatively) cheap access to water. As the Flint water crisis illustrates, however, public systems are by no means immune from issues of water interruption, water contamination, and high cost, one or all of which may send people looking for alternatives to the tap. Even before the switch to the Flint River, some Flint residents began to explore the possibility of drilling wells in their yard when several successive water rate increases made Flint water bills among the highest in the nation.[40] The deterioration of Flint's water quality in 2014 only further stimulated interest in this possibility. The option of having personal wells could bring peace of mind to residents whose faith in the public system has been lost. This, in turn, could invite a diversity of responses to the trauma of the water crisis. The overarching response to the crisis on the part of government and academic researchers has—for understandable reasons—been to try and restore resident faith in the public water system. Perhaps, however, a more just (as well as more realistic) response is to accept the fact that some residents are unlikely to ever trust the city government or water system again, and to seek to make possible, for them, a life where they can continue living within the city without having to rely as much on city water.

The viability of urban wells is often complicated in Flint and elsewhere by issues of groundwater contamination, and even in the absence of these issues, removing homes from the system can lead to higher costs and lower quality for those still on it, raising justice concerns. It is not inconceivable, however, that a shrinking city could identify areas where well-drilling could not only be accommodated but would help to facilitate decommissioning of infrastructure. If so, residents whose trust in the public system had been irrevocably lost due to a crisis like the one in Flint, or who for other reasons determined it to be in their interest to drink well water, could potentially do so in a city-sanctioned-and-assisted way.

Equity

Considering equity in a shrinking-city context also points to both opportunities and challenges. Morckel, for example, argues that reducing the size of Flint's water system will promote more equitable health outcomes. An overextended water system with insufficient demand leads to stagnation, which increases chances for bacteria to grow, among other water quality problems.[41] A pruned or downsized water system could ensure that water gets to people's taps quickly and safely. Another potential way in which rightsizing a water system can promote equity is by reducing high water rates that are especially burdensome to low-income residents. A 2016 water rate analysis found that water bills in Flint were high in part because the city has to maintain "96 feet of pipe per customer, versus the average of 83 feet per customer for peer communities."[42] The same analysis found that Flint was losing 50–60 percent of the water it purchased from the regional Great Lakes Water Authority because so much was lost through leaks on the way to the tap—a problem greatly exacerbated by having such a large amount of pipe to monitor and maintain. Once again, this problem translates into higher costs for residents who must help the city to recoup its losses in their water bill payments.

Other implications of shrinking or reconfiguring a water system may be harder to reconcile with the principle of equity. As mentioned above, one of Sadler et al.'s main concerns about rightsizing is that any relocation involved will be especially burdensome for people of color who form the majority in Flint's least-populated neighborhoods. For all of the issues these neighborhoods struggle with—blight, crime, dumping, etc.—they are also speckled with well-maintained homes and yards, cared for and invested in over the years by residents for whom the idea of a moderate-sized, free-standing home remains a core part of their American Dream. While the installation of new service lines allows this dream to persist into the indefinite future, calls for rightsizing potentially upend it by asking people to live in denser neighborhoods with smaller (or nonexistent) yards in the center city. The personal losses involved in relocation cannot necessarily be compensated for by providing people with access to a downtown lifestyle, especially when personal identities are shaped by attachment to the unique characteristics of places and neighborhoods. Giving up a house that has been in the family for generations, for example, or moving away from a community with which one has a long history, may weigh far more heavily in the balance than new dining and entertainment opportunities.

Conclusion

Whether the concerns we have raised above can be overcome is not for us to decide. Ultimately, it must be determined by shrinking-city residents themselves. It is likely that members of different communities will reach different conclusions about what justice looks like in their own community. It is also possible that what counts as a just shrinking city might change over time as these cities themselves change. So, rather than argue for a particular conception of the just shrinking city, we have instead sought to show that theorizing justice as an end goal in shrinking-city contexts could prove especially useful for moving these communities forward. The need, as we have identified it, is for careful, detailed proposals about what the just shrinking city might look like so that, within fair deliberative processes, residents can decide for themselves what is just.

We do wish to acknowledge one final challenge, however: that of "shrinking city" terminology itself. It may be that residents of shrinking cities are unlikely ever to come around to embracing the adjective "shrinking" to describe their hopes for the future. This may be in part because the term "shrinking city" risks contributing to what has been called "territorial stigma," giving the impression that a particular place is dying and dysfunctional.[43] Perhaps an entirely new term is necessary if the idea we have sought to capture in this chapter is ever to win public acceptance. Nevertheless, we do maintain that the idea itself, however articulated, has the potential to point us toward more constructive conversations about the urban crisis in Midwestern cities and beyond.

Acknowledgments

The authors are grateful for written comments from Ian Smith, Matt Ferkany, and an anonymous referee, as well as for conversations with Margi Dewar, Michael

Freeman, Justin Hollander, Shawn McElmurry, Victoria Morckel, Brent Ryan, and Rick Sadler. Any errors are the responsibility of the authors alone.

NOTES

1. Alan Mallach, *The Divided City: Poverty and Prosperity in Urban America* (Washington, DC: Island Press, 2018), 13.

2. Jon C. Teaford, *Cities of the Heartland: The Rise and Fall of the Industrial Midwest* (Bloomington: Indiana University Press, 1993), vii.

3. Mallach, *The Divided City*, 17.

4. Thomas J. Sugrue, *The Origins of the Urban Crisis: Race and Inequality in Postwar Detroit* (Princeton, NJ: Princeton University Press, 1996).

5. See Maxwell D. Hartt, "The Diversity of North American Shrinking Cities," *Urban Studies* 55, no. 13 (2017): 2946–2959, and Hartt, "The Elasticity of Shrinking Cities: An Analysis of Indicators," *Professional Geographer* 73, no. 2 (2019): 230–239. Although we refer to "shrinking cities" in a general way in this chapter, it must be kept in mind that not all shrinking cities are alike, and not all experience the full range of challenges we highlight here.

6. Hans Schlappa and William J. V. Neill, *From Crisis to Choice: Re-imagining the Future in Shrinking Cities* (Nancy, France: URBACT, 2013), https://uhra.herts.ac.uk/bitstream/handle/2299/19120/From_Crisis_to_Choice_June_2013.pdf?sequence=2.

7. Justin B. Hollander and Jeremy Németh, "The Bounds of Smart Decline: A Foundational Theory for Planning Shrinking Cities," *Housing Policy Debate* 21, no. 3 (2011): 349–367. See also Deborah Popper and Frank Popper, "Small Can Be Beautiful: Coming to Terms with Decline," *Planning* 68, no. 7 (2002): 20–23.

8. Justin B. Hollander, *An Ordinary City: Planning for Growth and Decline in New Bedford, Massachusetts* (London: Palgrave MacMillan, 2018), 21.

9. For a classic account of the history of planning, see Peter Hall, *Cities of Tomorrow: An Intellectual History of Urban Planning since 1880* (Malden, MA: Wiley-Blackwell, 2014).

10. Jeremy Németh et al., "Planning with Justice in Mind in a Shrinking Baltimore," *Journal of Urban Affairs* 42, no. 3 (2020): 351–370, 356. This paper refines and applies principles developed in Hollander and Németh, "The Bounds of Smart Decline."

11. Németh et al., "Planning with Justice in Mind," 3.

12. See David Harvey's *Social Justice in the City* (Athens: University of Georgia Press, 2009), 117. We do not deny that ideally justice in the city would consist of, to borrow David

Harvey's language, "a just distribution justly arrived at." But Harvey's phrasing itself implies that these elements of justice are, at some level, separable.

13. See Susan Fainstein's *The Just City* (Ithaca, NY: Cornell University Press, 2010), 66. Fainstein notes that "the equity outcomes of citizen deliberations are unpredictable" and that "planners can affect the character of deliberation and move participants toward a greater commitment to just outcomes."

14. See Brent D. Ryan, "Rightsizing Shrinking Cities: The Urban Design Dimension," in *The City after Abandonment*, ed. Margaret Dewar and June Manning Thomas (Philadelphia: University of Pennsylvania Press, 2012), 268–288. We acknowledge that the term "rightsizing" lacks a clear definition within the planning literature.

15. Mallach, *The Divided City*, 2.

16. See, for example, Ivonne Audirac's "Shrinking Cities: An Unfit Term for American Urban Policy?" *Cities* 75 (2018): 12–19, 18. She notes that "'Shrinking city,' 'urban shrinkage' or 'smart shrinkage' allusions are absent from the Detroit Future City Plan. Instead, [preferred is] repurposing much of the 20,000 acres of vacant land into ecological landscapes and obsolete industrial areas into 'live & make' districts."

17. See Brent D. Ryan and Shuqi Gao, "Plan Implementation Challenges in a Shrinking City: A Conformance Evaluation of Youngstown's (OH) Comprehensive Plan with a Subsequent Zoning Code," *Journal of the American Planning Association* 85, no. 4 (2019): 424–444.

18. Charles W. Mills, "'Ideal Theory' as Ideology," *Hypatia* 20, no. 3 (2005): 165–184.

19. See Andrew Light's "The Urban Blindspot in Environmental Ethics," *Environmental Politics* 10 (2001): 7–35, 35. As Light writes, "If environmental ethics is to fully embrace the urban, then it must describe the brown space of the city to be as important a locus of normative consideration as the green space."

20. For example, see Alastair S. Gunn's "Rethinking Communities: Environmental Ethics in an Urbanized World," *Environmental Ethics* 20 (Winter 1998): 341–360. As Gunn clarifies, the relevant sense in which a city would be natural is one where "human beings are able to thrive" (348). Naturalness, understood in this way, is presumably a normative concept.

21. Ingrid Leman Stefanovic, "In Search of the Natural City," in *The Natural City: Re-envisioning the Built Environment*, ed. Ingrid Leman Stefanovic and Stephen Bede Scharper (Toronto: University of Toronto Press, 2012), 24.

22. Roger J. H. King, "Environmental Ethics and the Built Environment," *Environmental Ethics* 22, no. 2 (2000): 115–131.

23. We recognize that, put this way, the concept of justice can get blurred with the concept

of the good. While we attempt to keep this blurring to a minimum in our own account, we note that such blurring is a widespread phenomenon within public discourse around justice and certain academic accounts of the just city.

24. "The Just City Index: Values Indicators," the Just City Lab, https://static1. squarespace.com/static/5b5dfb72697a9837b1f6751b/t/5b7d9e4221c67c1bee80 cbb7/1534959170568/JustCityIndex_Site.pdf.

25. See Toni L. Griffin and the Just City Lab from the Harvard University Graduate School of Design, *Design for a Just City*, 2019, https://static1.squarespace.com/ static/5b5dfb72697a9837b1f6751b/t/5e567c71e76e6430d0e49db3/1582726270369/ JustCityLab_Rotterdam.pdf. To be fair, the Just City Lab has a more fully developed discussion of the concept than we are able to summarize here.

26. Fainstein, *The Just City*, 3, 172–174.

27. Alan Mallach, "Comment on Hollander's 'The Bounds of Smart Decline: A Foundational Theory for Planning Shrinking Cities," *Housing Policy Debate* 21, no. 3 (2011): 369–375.

28. Benjamin J. Pauli, *Flint Fights Back: Environmental Justice and Democracy in the Flint Water Crisis* (Cambridge, MA: MIT Press, 2019).

29. For examples of this framing of the crisis, see Jacob Lederman, "Flint's Water Crisis Is No Accident. It's the Result of Years of Devastating Free-Market Reforms," *In These Times*, January 22, 2016; Jason Stanley, "The Emergency Manager: Strategic Racism, Technocracy, and the Poisoning of Flint's Children," *Good Society* 25, no. 1 (2016): 1–44; David Fasenfest, "A Neoliberal Response to an Urban Crisis: Emergency Management in Flint, MI," *Critical Sociology* 45, no. 1 (August 2017): 1–15.

30. *Imagine Flint: Master Plan for a Sustainable Flint: Summary of Goals and Objectives* (Houseal Lavigne Associates, October 28, 2013), 89.

31. David Streightfeld, "An Effort to Save Flint, Mich., by Shrinking It," *New York Times*, April 21, 2009.

32. The "not gonna happen" quote is found in Catherine Turner's *Small, Gritty, and Green: The Promise of America's Smaller Industrial Cities in a Low-Carbon World* (Cambridge, MA: MIT Press, 2012), 69, and the "not everyone" would "win" quote is found in Streightfeld, "An Effort to Save Flint, Mich."

33. Toward the end of the process, however, Flint was essentially overtaken by Newark, NJ, which neared completion of its LSL replacement program around the same time despite having begun a mere two years before. See Kevin Armstrong's piece on Newark's replacement of its LSLs: "'Hallelujah Moment': How This City Overcame Its Lead Crisis," *New York Times*, August 11, 2021.

34. "Lead Pipes Are Widespread and Used in Every State," National Resource Defense Council, https://www.nrdc.org/lead-pipes-widespread-used-every-state.

35. Victoria Morckel, "Why Flint (MI) Missed an Opportunity to 'Right Size' with Its Water Crisis," *Journal of the American Planning Association* 86, no. 3 (2020): 304–310, 305.

36. Morckel, "Why Flint (MI) Missed an Opportunity," 306.

37. Richard C. Sadler et al., "Right Sizing Flint's Infrastructure in the Wake of the Flint Water Crisis Would Constitute an Additional Environmental Injustice," *Journal of the American Planning Association* 87, no. 3 (2021): 424–432, 427.

38. Jacob Lederman, "The People's Plan? Participation and Post-Politics in Flint's Master Planning Process," *Critical Sociology* 45, no. 1 (2017): 1–17.

39. Scott Atkinson, "In Flint Neighborhood, Vacant Lots Will Soon Bear Fruit," *Next City*, July 20, 2017, https://nextcity.org/urbanist-news/entry/eastside-flint-greenspace-community-garden.

40. Kristin Longley, "Flint Water Rate Hikes Lead to Influx of Well Drilling Inquiries," *Mlive*, updated January 20, 2019, https://www.mlive.com/news/flint/2012/05/drill_baby_drill_flint_water_r.html.

41. Nancy Love, Richard Jackson, and Shawn P. McElmurry, "Water Stays in the Pipes Longer in Shrinking Cities—a Challenge for Public Health," *The Conversation*, May 24, 2019, https://theconversation.com/water-stays-in-the-pipes-longer-in-shrinking-cities-a-challenge-for-public-health-116119.

42. Raftelis Financial Consultants, Inc., *Flint Water Rate Analysis*, May 13, 2016, https://www.michigan.gov/documents/snyder/Flint_Rate_Analysis_Final_Raftelis_Report_May_13_2016_524463_7.pdf.

43. Audirac, "Shrinking Cities," 12.

SUGGESTED READINGS

Anderson, Michelle Wilde. "The New Minimal Cities." *Yale Law Journal* 123, no. 5 (2014): 1118–1222.

Clark, Anna. *The Poisoned City: Flint's Water and the American Urban Tragedy*. New York: Metropolitan Books, 2018.

Dewar, Margaret, and June Manning Thomas, eds. *The City after Abandonment*. Philadelphia: University of Pennsylvania Press, 2012.

Fainstein, Susan S. *The Just City*. Ithaca, NY: Cornell University Press, 2010.

Highsmith, Andrew R. *Demolition Means Progress: Flint, Michigan, and the Fate of the American*

Metropolis. Chicago: University of Chicago Press, 2015.

Mallach, Alan. *The Divided City: Poverty and Prosperity in Urban America.* Washington, DC: Island Press, 2018.

Pauli, Benjamin J. *Flint Fights Back: Environmental Justice and Democracy in the Flint Water Crisis.* Cambridge, MA: MIT Press, 2019.

Ryan, Brent D. *Design after Decline: How America Rebuilds Shrinking Cities.* Philadelphia: University of Pennsylvania Press, 2012.

Sugrue, Thomas J. *The Origins of the Urban Crisis: Race and Inequality in Postwar Detroit.* Princeton, NJ: Princeton University Press, 2005.

The Ethics of Women Water Warriors

#NoDAPL and an Indigenous Women's Environmental Ethic

Matthew Meyer and Heather Ann Moody

ndigenous peoples have existed within the confines of colonization and capitalism since first contact. As Temryss MacLean Lane (Lummi Nation) states:

> Indigenous women engage in the power of refusal on the frontline of ancestral homelands to disrupt colonial ideologies of knowledge and power. All of existence and ways of life are being threatened, compromised, and destroyed by capitalist values of imperialism that lack connection to land and healing modalities.[1]

Although facing continuing acts of oppression and dispossession, they stand in refusal through independence in asserting their sovereignty, treaty rights, and pressure to bring about political change. Indigenous women are at the heart of these acts due to their power and responsibilities within and to their communities. Standing Rock is the epitome of the power and responsibility of Indigenous women to utilize their ancestral knowledge to block the completion of the Dakota Access Pipeline. The Indigenous women present at Standing Rock, and elsewhere, are considered warriors. "Warriors of wellness in the face of violence. . . . Indigenous women define their missions of wellness and address direct challenges to community health they overcome in these spaces of refusal."[2] Looking at the foundation of health, wellness, and community through water protection at Standing Rock, we seek to address the intersection of Indigenous feminism and women's politics of resistance through the culmination of Indigenous knowledge, interconnectedness, nonviolent resistance, and maternal care. The main foundational concept is that Indigenous women are at the center of leadership, resistance, and protection, particularly concerning water; the hope, of course, is that people of all genders and backgrounds will continue to join them.

In this chapter, we will elaborate upon an ethics of environmental protection that is internal to the water protectors of Standing Rock, the women water protectors in particular. Our primary goal will be to elaborate this viewpoint from an internal perspective, while also acknowledging the shared features and the tensions that it creates with respect to Western perspectives on the environment and Western feminism. To that end, we begin by situating the context of water protection at Standing Rock and the #NoDAPL movement and then explain the Indigenous views of water, including the memorable phrase "water is life." While the term Indigenous may at first appear monolithic, our goal is to explain the cultural context of the importance of water among the Lakota and Anishinaabe as well as to show that the most central concerns about water are shared by many other nations around the world. We also discuss the importance of Indigenous knowledge, frequently referred to as traditional ecological knowledge (TEK), to the issue of water protection. Then, we compare and bridge the Indigenous ways of understanding the environment to two views of the environment in Western environmental ethics: biocentrism, or the belief that anything alive has inherent moral value, and ecocentrism, or the belief that natural systems have inherent moral value as they support living beings. The next section addresses the concept of interconnectedness through relationships—those to each other, the land, and to past and present generations—leading into a discussion of reciprocal responsibilities. Incorporating this knowledge, we see that there are three key features in an Indigenous women's ethics of water protection: interconnectedness, the focus on Indigenous knowledge, and the focus on the maternal and motherhood as a special area of responsibility toward water. For this last point, which recurs throughout the chapter, the following section focuses specifically on the importance of the maternal with comparisons to Sara Ruddick's *Maternal Thinking*, specifically her ideas about a "Women's Politics of Resistance." We then address nonviolent Indigenous Resistance regarding water protection. This leads us back to revisit

Standing Rock, and we come full circle to the final section, which addresses the universalization of water protection. This is done to demonstrate that even though many Indigenous views about the environment can be grounded in cosmologies and stories that may not ring true to all of us, the tactics of Indigenous water protection share similar features to other forms of resistance, and can be, if not universalized, expanded upon for other environmental and political concerns.

There is an important point to note before we begin elaborating on an Indigenous women's environmental ethic. We will be describing positions that, if one is reading this solely from a Western perspective, may not jibe with current biological science, or in other cases might strain against entrenched (White) feminist critique. On the one hand, we will do our best to acknowledge those tensions and defend the Indigenous positions. On the other hand, a defense should not necessarily be required. To hold Western science or Western feminism as the standard-bearer of all things "knowledge-related" is precisely the problem. Doing so is what allows one to think: "Well, all of these Indigenous ways of thinking would be helpful in solving environmental problems, but we really do have to get _____ (fill in: water, energy, people fed, etc.) somehow." What this fails to acknowledge is that there are other ways of living, and thinking and claiming that "it would be nice to think otherwise, but . . ." is reverting to default thinking, the status quo. It could be the case, in fact, that it is not that the Indigenous way of thinking is an unhelpful mismatch to Western ways of living, but rather that the Western ways of living and knowing are themselves limited and problematic.[3] In short, we do our best to present, compare, and contrast an Indigenous view to a Western view, while also acknowledging that the former's differences from the latter should not be seen as weaknesses or inferiorities but rather as strengths and sometimes even as complementarities.

Throughout this chapter we focus on two groups of Indigenous peoples. One group is the Lakota, who are specifically related to the protections at Standing Rock, located on the border of North and South Dakota in the central part of the two states. The second group is the Anishinaabe, located primarily in northern Wisconsin and Minnesota and southern Ontario, as they have been very active with water protection in the Midwest as well, including prominent leaders bridging Indigenous and non-Indigenous people in the protection of water. The terminology utilized in referring to Indigenous peoples is complex. For clarification, we are using the terms used by the peoples themselves throughout the chapter. Therefore, you may see the terms Sioux, Lakota, and Dakota used depending on the source, even though these terms often refer to overlapping peoples. In general, we are respecting that the term Sioux is a term that was imposed on the Lakota, Dakota, and Nakota peoples, and therefore when allowed, we will refer specifically to the Lakota peoples. In reference to the Anishinaabe, we recognize that there are different spellings of this term, and that the term itself encompasses the culturally related peoples of the Ojibwe, Potawatomi, and Odawa. Again, where necessary, we will refer specifically to the Ojibwe depending on the source. Ironically, these two Indigenous peoples (Lakota and Anishinaabe) were historically enemies but are seen today working together for not only the protection of water but of the natural environment we all live in. Their relationships are representative of those that need to be built to protect the environment for all living beings.

Water Protection at Standing Rock—#NoDAPL

Nick Estes (Lower Brule Sioux Tribe) reflects on Standing Rock, "From late summer of 2016 to the winter of 2017, more than three hundred Native nations planted their flags in solidarity at Oceti Sakowin Camp" to protect the water from the construction of the Dakota Access Pipeline.[4] With tradition and responsibility as the foundation, women at Standing Rock were actively protecting the land and water through acts of nonviolence, in the face of violence. As Temryss MacLean Lane points out, the movement of women began in April 2016 when LaDonna Brave Bull Allard (Standing Rock Sioux Tribe) "put out a call to her tribal relatives" to help her stop the construction of the Dakota Access Pipeline. After Brave Bull Allard's initial call to action, answered primarily by youth, the call to action went beyond the Standing Rock people. They utilized "social media as the modern-day smoke signal, asking all relatives, Native and non-Native, to come 'stand with Standing Rock' in protection of the water."[5]

Throughout history, on a federal and state level, the Lakota people have had their lands infringed upon involuntarily through acts of colonialism. The construction of the Dakota Access Pipeline (frequently referred to simply as DAPL) is a modern example of how colonialism continues to impact the Lakota people. The gathering of Tribal Nations at Oceti Sakowin (literally "Seven Council Fires," the proper name for the Sioux, and made up of seven peoples) was intended to be an anticolonial movement to bring awareness to the impact of the Dakota Access Pipeline on not just Indigenous peoples but all people, as well as demonstrate in a peaceful manner their opposition to the pipeline. The biggest concern with the pipeline was the impact to the major waterways, in particular the freshwater that is produced by the nation's largest watershed, the Missouri-Mississippi, which might be contaminated by a leak. "The pipeline will carry half a million barrels of heavy crude oil a day across four states, . . . under the Missouri River twice, and under the Mississippi River to refineries in Illinois and the Gulf of Mexico."[6] Although Energy Transfer Partners, the owners of the Dakota Access Pipeline, claimed the pipeline would be safe, history has proven that all pipelines leak. And in fact, the Dakota Access Pipeline did leak five times in 2017: twice in North Dakota, once in South Dakota, once in Illinois, and once in Iowa.[7] With this in mind, it was imperative that the water be protected, as we all need water.

The movement surrounded the Lakota concept and signature statement of the movement, *Mni Wiconi* (Water is Life). According to Dhillon and Estes,

Mni Sose [the Missouri River] is not a thing that is quantifiable according to possessive logics. Mni Sose is a relative: the Mni Oyate, the Water Nation. She is alive. Nothing owns her. Thus, the popular Lakotayapi assertion "Mni Wiconi"— *water is life* or, more accurately, *water is alive*. You do not sell your relative. . . . To be a good relative mandates protecting Mni Oyate from the DAPL's inevitable contamination.[8]

The #NoDAPL movement at Standing Rock is one that is grounded at the intersection of Indigenous and non-Indigenous environmental ethics and feminisms, particularly as they relate to women's responsibility for and maternal thinking about the environment. While very little has been written about the intersection of Indigenous and non-Indigenous conceptions of maternal ethics and the women water protectors at Standing Rock, what follows is an attempt to bridge this gap by offering an Indigenous and non-Indigenous account of maternal thinking as an environmental ethic at work through the act of water protection.

While the larger picture of what's "wrong" with the Dakota Access Pipeline is complicated (for example, it promotes continued use of fossil fuels, it enacts what amounts to eminent domain for the purpose of fossil fuel company profits, etc.) the main environmental issue for the Standing Rock Sioux Tribe was the pipeline's crossing of the Cannonball and Missouri Rivers, which are both a sacred site and a source of drinking water for the people.[9] One point of symbolic note here is that the Mni Sose, the Missouri River, is also known as the Mother River.[10] Thus, the Missouri River forms the foundation of several interconnected forms of life: biological life, spiritual life, and the life-giving capacity of the mother, as we will see in the following narratives.

Mni wiconi ("water is life") is a common phrase echoing throughout Indian Country during the resistance to the Dakota Access Pipeline in North Dakota. The term "water protector" is frequently used to describe all people who are against recent pipeline expansions. However, the role of water protector, particularly among women, is not new. Indigenous women have held the responsibility for protecting the land, her resources, and thus the people, for generations. Debra Topping (Fond du Lac Ojibwe) puts it simply: "I need to protect this . . . for our future generations, because that's who we are as a people."[11] This sentiment is furthered with the earlier statement that Indigenous women are "warriors of wellness in the face of violence."[12] As we suggest throughout the chapter, women have always had the cultural responsibility to be "warriors of wellness," which naturally intersects with water protection. We are interested in the Indigenous women water warriors, or water protectors, at Standing Rock because they are at the intersection of two models for environmental protection: Indigenous feminism and a women's politics of resistance.

By drawing from works that address women as resisters as well as women as water protectors, we will be defining and identifying key themes that make the movement an Indigenous environmental ethics of resistance, with a focus on maternal thinking. Specifically, we identify the principles of knowledge, awareness of interconnection and reciprocal responsibility, and a women's politics of resistance—including a focus on maternal care—as central to the #NoDAPL movement at Standing Rock. In the conclusion, we will revisit Standing Rock as a model of an Indigenous women's environmental ethics of care and how we can all be water protectors.

Indigenous Views of Water

In order to fully understand the concept of water protection and water warriors, it is vital to understand the Indigenous view of water and how that extends beyond a Western scientific view of water. Water is connected to life beyond the biological needs of living beings. Ardith Walkem (Nlaka'pamux Nation) states that "water is the lifeblood of the land and of the [I]ndigenous peoples and cultures that rely upon it." Within Indigenous cultures water is intertwined with culture including how to treat water. "Water is a living thing, a spiritual entity with 'life-giving' forces, which comes with certain duties and responsibilities to ensure that it is respected, protected, and nurtured." At the center of this care are women. As life-givers themselves, Indigenous women are tasked with caring for water through various means, including ceremonies and songs. Due to these responsibilities towards water, Indigenous women have often been referred to as "Keepers of the Water," "Carriers of the Water," and "Water Walkers."[13] Water, therefore, moves beyond a Western scientific view of water as necessary for survival and into the wholistic view of water within culture for Indigenous peoples serving as a central component for ceremonies, medicines, and healing.

For the Lakota, the phrase *mni wiconi*, referenced above as "water is life" and which became a signifying phrase for the #NoDAPL movement, appears simple and straightforward. However, the terminology is not so simple and upon examination represents a fuller understanding of the relationship between humans and water. Tiokasin Ghosthorse, a Cheyenne River Lakota activist and international advocate, explains that *mni* goes beyond the simple translation of "water."

> The *Ni* (nee) is "life" and could also mean "mother's milk" or a "mother's breast." This is where the "M" of *Mni* becomes translatable as "you and me" but also becomes a little more understandable if we say *Mni* is "you and me of that which carries or causes feeling with another through itself." Like a mother who is the carrier of Water, *Mni* is an action of living."[14]

With this cultural context of understanding the term *Mni*, we also get an understanding of not only "water is life" but also of "water is alive." Ghosthorse further explains the cultural context of water within the creation story of the Lakota people as being "a First Consciousness bestowed upon Mother Earth," which he explains as "the awareness of the movement that sustains life in a continuum."[15] The creation story references water as "blue blood" that courses throughout Mother Earth and thus is foundational to all life, thus reflecting the deep relationship between the people and water and the need to protect the water.

A similar relationship is seen among the Anishinaabe people as well, which emphasizes the specific relationship of women to water.

The Earth is said to be a woman. In this way it is understood that woman preceded man on the Earth. She is called Mother Earth because from Her come all living things. Water is Her life blood. It flows through Her, nourishes Her, and purifies Her.[16]

It is clear that women are heavily connected to water in both Lakota and Anishinaabe cultures. Additionally, it is also clear that water is considered a living entity. The Anishinaabe clearly emphasize this within their creation story, as seen above. Similar to what we have seen with the Lakota, water is central to the Anishinaabe as a source and supporter of life, beyond that of biological function.

Water mediates interactions among many living beings on the [E]arth and is considered a relative with responsibilities to give and support life. Bodies of water are considered to have their own unique personalities. Humans have special responsibilities to respect and care for water and to encourage its life-giving force.[17]

If one thinks within a Western mindset, it may be difficult to understand that water is living. To begin with, there is a widely accepted definition of what constitutes life that includes seven characteristics: organization, homeostasis, metabolism, growth, response, reproduction, and adaptation.[18] Clearly water itself does not possess these features. What is notable, though, is that many of these characteristics *require* water for their operation. Water aids with homeostasis (in the case of animals, it helps regulate body temperature) and with metabolism. Furthermore, water is considered a "universal solvent," one that "helps cells transport and use substances like oxygen or nutrients."[19] Thus it can be argued that in a typical biological account, while water is not itself alive, it is a requirement for life. In acknowledging this, we can establish a relationship of dependence such that without clean, soluble water, we cannot have life as we know it. Still, in most Indigenous understandings of water, this relationship of water to life can be examined even more closely. In many such views, water is seen as a "relative," much as Mother Earth is a relative. Accordingly, for most Indigenous peoples, there is a responsibility to care for water, as a reciprocal relationship to the idea that water cares for us; this responsibility primarily rests in the hands of Indigenous women.

Although we are focusing on Lakota and Anishinaabe peoples, there are general concepts that are universal to most Indigenous peoples throughout North America. The concept that the Earth is a living entity is vital to understanding the relationships Indigenous peoples have to their natural environments. As more specifically addressed above, the Earth being referred to as "Mother Earth" represents a life-giving body to the people. Although the specifics of the origins of why water is important to Indigenous peoples varies in detail and is location specific, there is no dominion over the Earth within Indigenous worldviews, that is universal.

A Western Environmental Ethics Perspective

From a strictly Western biological standpoint the claim that water is alive is problematic, as we have seen. From a standpoint of poetics, less so: the flow of water mimics animate beings. In fact, if it is not alive, it is one of the few substances (maybe along with fire and wind) that acts *as if* it were alive. Furthermore, it is clear that "alive" in the previous accounts contains two claims: one, that water itself flows and moves, and in that way is analogous to something alive. But second, and more importantly, water gives and sustains life. This second claim can be taken "scientifically" in that all living things need water to survive. But in a poetic sense, the distinctly human female aspects of giving life—both the menstrual cycle and the amniotic fluid ("my water broke")—share the flowing aspect of water. All that said, we need not take the claim that "water is alive" scientifically, but instead as a "regulative idea" in the Kantian sense. In this Kantian sense, treating water as if it were alive (even if such a claim is counter to current scientific knowledge) reminds us of its importance in sustaining life and can guide our behavior. We can view the claim from the much more palatable scientific claim that water is required for living beings, that it promotes growth and sustains living beings.

The viewpoint that water is required for life is not only scientifically acceptable but puts water squarely within the realm of being understood as a morally considerable entity according to two closely related frameworks for environmental ethics. First one word on what we mean by "moral considerability": in "On Being Morally Considerable," Kenneth Goodpaster introduces what he calls the "life principle."[20] In short, all *living* beings deserve moral considerability. "Moral considerability" is the idea that the value of the being in question ought to be considered when weighing moral decisions. As Goodpaster himself notes, "moral considerability" is not a one-size-fits-all proposition: even though both goats and humans might deserve moral consideration (based on the life principle), they may not be said to have equal moral significance. In other words, we may still give more moral weight to one species than to another.

Because of its focus on the life principle, Goodpaster's view is biocentric: it focuses on life as the central feature of moral value. At first glance, Goodpaster's biocentric view might seem to exclude the considerability of something like water. However, as he notes in response to the objection that the "life principle" might have to include "the biosystem as a whole": "There is some evidence that the biosystem as a whole exhibits behavior approximating the definition [of life] sketched above, and I see no reason to deny it moral considerability on that account."[21] This biosystem as a whole, as we will now discuss, most surely includes water.

The other view that extends moral considerability to water, held by Aldo Leopold and others, is now called ecocentrism. Let us clarify what we mean by this. There are two views of Leopold's land ethic, and the moral considerability of something like water could rest on which view we take. Each of these views are outlined by J. Baird Callicott in different essays: "Animal Liberation: A Triangular Affair" and "The Conceptual Foundations of the Land Ethic." In "Animal Liberation" the view that is put forward is one of holism—namely, that the continued existence of the biological community is more important than the continued existence of any given member of that community. Indeed, this seems to accord with Leopold's own definition of the land ethic—namely, that things are right to the extent that they "preserve the integrity, stability, and beauty of the biotic community."[22] Accordingly, hunting for the sake of culling an overpopulated area of deer would be acceptable, whereas preserving an invasive species simply because of its aesthetic charms would not. Considering this point, Callicott argues: "Animals of those species, which, like the honey bees, *function in ways critically important to the economy of nature*, moreover, would be granted a greater claim to moral attention than psychologically more complex and sensitive ones, say rabbits and moles, which seem to be plentiful, globally distributed, reproductively efficient, and only routinely integrated into the natural economy."[23] While the reference here is to species, the claim that water "functions in ways critically important to the economy of nature" seems obvious. Furthermore, calling this view "ethical holism," Callicott writes, working backwards from the logic above:

> One cannot affect the system as a whole without affecting at least some of its components. An environmental ethic which takes as its *summum bonum* the integrity,

stability, and beauty of the biotic community is not conferring moral standing on something *else* besides plants, animals, soils, and waters.[24]

So even if we function on holism, this first reading still accords more moral attention to those things that are central to the continued existence of the biotic community, and one of these things is water. It seems to us that this would be just as true for individual instances of watersheds, such as the Missouri River ecosystem, as it would for "water in general."

The second view of Leopold's land ethic, given by Callicott in "The Conceptual Foundations of the Land Ethic," focuses more on some of Leopold's "minor" principles, such as taking the "trend of evolution" to "diversify the biota" as an operational principle. Callicott here lists the "commandments" that emerge from this view of the land ethic: "Thou shall not extirpate or render species extinct; thou shalt exercise great caution in introducing exotic and domestic species into local ecosystems, in extracting energy from the soil and releasing it in the biota and *in damming or polluting the water sources*; and thou shalt be especially solicitous of predatory birds and mammals."[25] In other words, even if we move the land ethic away from holism and toward the preservation of biodiversity, water *still* occupies a central place in our ethical consideration.[26]

By either of these logics of the land ethic, altering a water source that gravely affects the living beings that depend upon it is the same as directly affecting the beings themselves. To this end we can consult Leopold's own words: Aldo Leopold in his famous essay "The Land Ethic" puts water, land, and soil at the base of his "land pyramid." Being at the base of the pyramid is significant, because it means that all other living beings depend upon those entities. "Water is life" means that without water, or with contaminated water, the ways of life of many living beings will be altered or ended.

An additional insight of Leopold's is relevant here. In the same essay he discusses what is called the "The A-B Cleavage." By this he means that we can demarcate two distinct "attitudes," what we might now call "ways of seeing" the land. Discussing agriculture, he says: "One group (A) regards the land as soil, and its function as commodity-production. Another group (B) regards the land as biota, and its function as something broader." Group B understands the land pyramid and its accompanying circuit of energy. He points out specifically of water: "Waters, like soil, are part of the energy circuit. Industry [group A], by polluting waters or obstructing them with dams, may exclude the plants and animals necessary to keep energy in circulation."[27] Goodpaster's view of the bio-system—life-sustaining as it is—and Leopold's view of biota as an energy circuit both arrive at the same conclusion, if by different paths. Namely, water deserves moral consideration—and has a considerable moral significance—because of its central place in sustaining all other living things.

The relation of interdependence exhibited in both Western environmental views of biocentrism and ecocentrism gets at another central point of Indigenous environmental views: that the only proper understanding of beings is the understanding of their relations to each other. They are views grounded in interconnectedness and interdependence. The reason, at bottom, that it does not matter whether water itself as an entity is determined to meet the scientific criteria for being alive in itself is because there is no "in itself" according to the Indigenous/biocentric/ecocentric/ecological view. The only thing that would matter is that with the depletion or contamination of water, many other living beings would most certainly be affected and/or perish.

Indigenous Ways of Knowing

Robin Wall Kimmerer (Citizen Potawatomi Nation), author of *Braiding Sweetgrass: Indigenous Wisdom, Scientific Knowledge, and the Teachings of Plants*, describes the distinctions between Western mentalities and Indigenous ways of knowing:

> In the Western tradition there is a recognized hierarchy of beings, with, of course, the human being on top—the pinnacle of evolution, the darling of Creation—and the plants at the bottom. But in Native ways of knowing, human people are often referred to as "the younger brothers of Creation." We say that humans have the least experience with how to live and thus the most to learn—we must look to our teachers among the other species for guidance.[28]

Braiding Sweetgrass is a prime example of traditional ecological knowledge (TEK) bridging the worlds of Indigenous and Western science. Yet, TEK rarely gets acknowledgment for its contributions to Western science. TEK "refers to the knowledge base acquired by [I]ndigenous and local peoples over many hundreds of years through direct contact with the environment."[29] Although the decisions of many Indigenous communities are informed by TEK, it is important to note that there is no one dominant way of knowing about the natural world or human-nature relationships and therefore no one universal definition of TEK. In addition to no one universal definition, there is no one way to apply TEK. Rather, each community conceptualizes and applies TEK in ways that are applicable to their specific cultures and environments. TEK is location-specific and, as we have demonstrated with the Lakota and Anishinaabe's relationship with water, focuses on the relationships between landscapes, time, and lifeways.

The concept of ethnoscience builds on the concept of relationships with the environment by including culture within science, rather than excluding cultural context. Ethnoscience evolved as a term in the 1960s to encompass the wholistic look at explaining the world around us. Rather than rely on strictly Western science, ethnoscience incorporates the viewpoints and culture of peoples who have accumulated thousands of years of TEK. Indigenous peoples "maintained a wholistic understanding of the natural world until forced to change by the intrusions of Western civilization."[30] As with the impact of colonization and Western movement within North America, traditional ways of knowing were pushed aside and replaced with Western ways of knowing as the norm and superior ways of living. This replacement of TEK with a Western perspective also applies to water protection in the context of Standing Rock. The element of relationships within TEK is vital to water protection. Often within the context of Western science and public policy, the relationship of humans to water is purely one of biology or commodity and does not account for the recognition of water as living in a reciprocal relationship with the living entities around it, resulting in a water as commodity/resource versus water as relationship situation.

It is important to note that TEK is sometimes referred to as simply "traditional" ways of knowing or "ecological knowledge," both of which tend to be broad and vague. As Berkes indicates, "*traditional* usually refers to cultural continuity transmitted in the form of social attitudes, beliefs, principles and conventions of behaviour and practice derived from historical experience." Additionally, when looking at the definition of ecological knowledge, it is frequently associated with ecology, which is usually considered to be a branch of biology within Western science. The issue then associated with "ecological knowledge" lies in the fact that "most traditional peoples are not scientists."[31] There has been some movement with a preference for "Indigenous ecological knowledge" that emphasizes Indigenous peoples specifically. No matter what terminology is used, TEK remains a wholistic approach to science and must be recognized as such. Frequently those within Western science view TEK as anti-science and subjective, resulting in the perception that TEK is inferior to Western science and emphasizing the notion that "science" must be viewed in one particular manner, which is Western science. That said, TEK is most valuable when it is viewed not as an "alternate" to Western science, but as complementary to it.

Incorporating Other Ways of Knowing

Kyle Powys Whyte, a member of the Citizen Potawatomi Nation, addresses the connection between Indigenous ways of knowing and the #NoDAPL movement:

> While appearing as direct action or a standoff, many Indigenous persons whose work created the #NoDAPL movement say it is really ceremony, prayer, and water protection. The meanings of English-language expressions, such as "ceremony" and "water is life," arise from time-tested Indigenous knowledges that prescribe respectful moral relations with water and other nonhuman beings and entities as vital for securing human safety and wellness.[32]

As an environmental ethical principle, the principle of knowledge refers to the imperative to gather all relevant sources of knowledge regarding people's interaction with the environment prior to making a policy decision. Importantly, this should include TEK of an area as a resource.[33] The women's water protector movement employs knowledge that falls into three categories: TEK, historical knowledge, and Western ecological knowledge. Each one of these types of knowledge serves a specific purpose in the discourse of the water protectors. TEK is a part of the discourse because of its long-standing role as the primary and more authentic mode of knowledge for Indigenous people. These knowledges and technologies have the distinction of having made it possible for Indigenous people to live in concert with the environment for thousands of years. By contrast, the past knowledge of Western and European societies, which often is founded on a Christian frame of having dominion over the earth, fails to sustain resources and instead extracts them to use them up.[34] Nonetheless, protectors also turn to historical and scientific accounts of environmental injustice primarily to "speak the language" of the policymakers, and everyday people, they need to persuade.

By historical knowledge we mean specifically a more thorough accounting of human interaction with a specific place and/or with other people concerning a place. Lastly, by Western ecological knowledge we mean to focus on an account of interactions between organisms and their abiotic environment. Often these latter two accounts overlap to the extent that a history of a people and place may include accounts of physical, chemical, or ecological changes; thus it becomes a historical account through a scientific frame—that is, environmental history. We will now give an example of each of these types of knowledge as they relate to the protection of the *Mni Sose*.

TEK: Indigenous Knowing

While numerous examples abound of the appeal to Indigenous knowledge in the context of #NoDAPL and water protection, one example of the way in which knowledge of the environment can be woven into accounts of personal development is in the account of "river knowledge" from Lanniko Lee (Cheyenne River, Lakota):

> We linked river knowledge and river power to the knowledge and power in our human lives. Late one summer evening when my grandmother and I were down at the river getting water, a light breeze started the cottonwood leaves chattering. ... She told me to stop and listen. ... "Listen to the cottonwood women talking," she said to me. I asked her what they were saying, and she said that I should always remember that I am like the river and that one day I would flow and become a woman. ... It was then that I learned about how the seeds of cottonwood trees are imbedded into the soft river soil by the swift moving waters that flooding causes in the spring run-off. I learned about the gift of life, and she made clear to me the Lakota saying, "Our children are sacred," with her explanation of how young cottonwood trees were the gifts of floods. I learned then to respect those powerful changes and appreciate our connection to the river, the trees, and the land.[35]

In this account, Lee recounts how her grandmother's narrative likened seeds to children, thereby expanding what is meant by "Our children are sacred." In this instance, anything that is a continuance of life is valuable offspring, as opposed to favoring human offspring. What's more, the natural processes of women's life-giving capacities—menstruation—is likened to the life-giving, flowing water of the river. Lee's account gives meaning to the idea of the river being a "mother," especially the Missouri (at least prior to its being polluted from agricultural runoff) being called the Mother River.

Historical Knowing

Winona LaDuke (White Earth Ojibwe) reminds us that the Missouri River is officially part of the last treaty regarding that land—"the fertile crescent of North America"—until the U.S. government flat out took the same land "as retaliation against Sitting Bull's victory at the battle of Little Big Horn" in 1877. As if that weren't bad enough, the 1944 Pick Sloan dams "flooded out the Missouri River tribes, taking the best bottom lands from the Mandan, Hidasta, and Arikara, the Lakota and Dakota. Over 200,000 acres on the Standing Rock and Cheyenne River reservations in South Dakota were flooded by the Oahe Dam itself, forcing not only relocation but a loss of the Lakota world." The reservoir "eliminated 90 percent of timber and 75 percent of wildlife on the reservations." Fast forward to the Dakota Access Pipeline, whose "northern portion was moved away from the water supply of Bismarck, into the watershed of Standing Rock." It would "span 200 water crossings, and in North Dakota alone, would pass through 33 historical and archeological sites."[36]

It is important to know that one of the original routes proposed for the Dakota Access Pipeline ran ten miles north of Bismarck, North Dakota—a city that is predominantly white. According to the *Bismarck Tribune*, "one reason the route was moved was because it threatened Bismarck's water supply."[37] As the paper notes:

> The U.S. Army Corps of Engineers evaluated the Bismarck route and concluded it was not a viable option for many reasons. One reason mentioned in the agency's environmental assessment is the proximity to wellhead source water protection areas that are avoided to protect municipal water supply wells.

Perhaps, by a utilitarian calculus, one could argue that *more* people could be affected by a spill on the northern route. That said, what should amaze us is that, according to Karen van Fossman, "we never even in Bismarck had to make an objection. The pathway was moved away from our drinking supply without our even needing to go to a meeting or write a letter."[38] One could argue, "Well, the pipeline needs to go somewhere!" But to do so is to be rather naive in our understanding of what is "necessary." It shows, in fact, that our tendency is by default to believe whatever Big Oil says is required, when the only reason it is "required" for them is to make additional profit. We agree with the Reverend Jesse Jackson, who said that the movement of the pipeline constitutes "the ripest case of environmental racism I have seen in a long time."[39] Thus, when it comes to the respect of the Lakota on the part of both Energy Transfer and the state and federal governments, the message is the same as it has always been: the use of land by settlers—even for environmentally detrimental purposes—is to be prioritized over natural, spiritual, and even treaty claims of Indigenous people. "The position of the Standing Rock Sioux Tribe is that the Dakota Access Pipeline violates Article II of the [1851] Fort Laramie Treaty, which guarantees the 'undisturbed use and occupation' of reservation lands surrounding the proposed location of the pipeline."[40] Importantly, this most recent violation of treaty rights can only be seen in the fullness of its injustice in virtue of the historical context.

Western Ecological Knowing

Lastly, Indigenous groups fighting pollution and environmental injustice recognize the relevance of Western ecological knowledge. In *All Our Relations*, Winona LaDuke discusses Katsi Cook's keen awareness of the damage done by toxins in the body in the "Akwesasne's standoff with General Motors." According to Cook (Mohawk Nation at Akwesasne) "women are the first environment."[41] When GM dumped PCB and materials containing heavy metals into the groundwater, it was the Mohawks and the Haudenosaunee who were affected. LaDuke quotes Cook directly:

> We accumulate toxic chemicals like PCBs, DDT, Mirex, HCBs, etc., dumped into the waters by various industries. They are stored in our body fat and are excreted primarily through breast milk. What that means is that through our own breast milk, our sacred natural link to our babies, they stand the chance of getting concentrated dosages.[42]

Cook's scientific awareness also highlights two key themes of Indigenous aware-ness, namely, a specific concern for the health of mothers as the givers of life, and the awareness of the interconnectedness of all things.

Lucid Knowing

As one can see, two of the above examples speak directly about motherhood as a central value and imperative for the application of knowledge in an environmental ethic. In her book *Maternal Thinking: Toward a Politics of Peace*, Sara Ruddick sees maternal thinking as an ethic that employs two main features: the "primacy of bodily life and the connectedness of self and other."[43] In a key section of that same work entitled "A Women's Politics of Resistance," Ruddick lays out three features that make maternal thinking particularly apt for nonviolent resistance: a desire for lucid knowledge, an awareness of the pain of others, and holding to a stubborn decision in the face of opposing forces.[44]

Ruddick elaborates upon the feature of "lucid knowledge" by applying the idea to the Madres of Argentina. Lucid knowledge refers to the desire to gain knowledge about and make public the atrocities committed against loved ones and homelands. The Madres of the Plaza de Mayo were survivors of government violence perpetrated against their loved ones (almost all male loved ones) by Argentina's brutal authoritarian government. Over 30,000 of their husbands, sons, and brothers were "disappeared" by the authoritarian Junta government between 1976 and 1983. The Madres courageously stood up to the brutal government violence with nonviolent, silent protests on the Plaza de Mayo beginning April 30, 1977. They protested right outside the building that housed the leaders who had disappeared their loved ones. Initially, only fourteen courageous women showed up. By October, the number grew to 237.[45] Soon thousands joined. The mothers held signs, wore placards around their necks with pictures of their loved ones, and held candles. Their lucid knowledge came in the form of demanding the whereabouts of their loved ones, and in their continued persistence even after the dictatorship fell. In recent years, this has even turned to DNA tests to identify bodies in mass graves.[46] Lucid knowledge builds from the desire to make known the acts of violence perpetrated on the mothers' loved ones through making themselves seen: the Madres stood in for their lost loved ones by wearing pictures on T-shirts, through making tapestries and *arpilleras* (quilts), and telling their stories to anyone who would listen.[47] Many Indigenous water protectors exhibit the same feature of lucid knowledge through sharing their stories through dance and through prayer. Increasingly, activists are making their stories known through more Western forms of storytelling, such as the documentary *Awake: A Dream at Standing Rock*.[48]

By having an "awareness of the pain of others" Ruddick means that in maternal thinking, the pain of personal loss can be extended beyond one's own situation to feel for others, to fight against violence and oppression. While Ruddick uses the example of how the Madres of Argentina came together through shared loss, one could also see groups such as Mothers Against Drunk Driving, or even Cindy Sheehan protesting the second Iraq War as examples of this awareness. In all of these cases, attending to the pain of others was not going to bring back their own loved ones. But through a combination of the awareness of their own and each other's pain, and making the dire facts of that particular situation known (be it the Argentinian Junta, or Chile's Pinochet, or drunk driving, or an unjust war), these women rightly believe that they can do something to prevent others from experiencing this pain in the future.

By making a "stubborn decision" Ruddick means overcoming one's insecurities and even, possibly, one's role in society to make a stand for a just cause. In the particular configuration of Chilean and Argentinian society in the late 1970s and early 1980s, women were still not seen as "public" figures. And yet, the Madres consistently stood up to protest the authoritarian governments that were the source of their pain. Similarly, the women at Standing Rock were facing intimidation and brutality from the politically and financially stronger petro-state apparatus, and yet continued to peacefully dance, pray, and occupy the sacred space that the apparatus was trying to take from them.

There are two potential criticisms of Ruddick's work, and we would like to address those here. The first, like the accusations leveled against Indigenous feminists, is that to treat women in terms of the maternal is to be essentializing.[49] It is to overdetermine the existence of women by implying that they are essentially mothers. The second is that the work of maternal thinking is ethnocentric, in that Ruddick generalizes features of maternal thinking from her position of being a white, Protestant, college-educated mother.[50]

The first criticism can be addressed by pointing out that there is a key difference between saying, "All women are meant to be mothers," and saying, "Thinking like a mother includes the following features." Ruddick is arguing the latter. In fact, she goes so far as to say that the role of "mother" is "sex-neutral." In the preface to the 1995 edition of *Maternal Thinking: Toward a Politics of Peace*, Ruddick explains: "Anyone who commits her or himself to responding to children's demands, and makes the work of response a considerable part of her or his life, is a mother."[51] For Ruddick, a mother is not a "fixed biological" or legal entity, but rather defined by the work someone does. Thus, mothering can extend well beyond the biological realm of the female and need not be equated with it. For that reason, the essentialist critique is unfounded. The second criticism of ethnocentrism is undermined by the very fact that Ruddick includes the experiences of mothers from all walks of life in *Maternal Thinking* and bases her articulation of nonviolent feminist politics on the experiences of mothers in Chile and Argentina.[52]

There is another aspect worth pointing out about the "essentialism" that the focus on motherhood brings up in feminist debates from the perspective of Indigenous feminism. Indigenous feminism is grounded in ways of life that extend back thousands of years—not just from second wave feminism. We bring this up for two reasons. One, motherhood is central to most forms of Indigenous thinking. Verna St. Denis (Métis/Okemasis) and Kim Anderson (Cree/Métis) are two Indigenous scholars who explain this phenomenon. According to St. Denis, there is a "distinct valorization of maternity and motherhood in Aboriginal cultures."[53] However, unlike in the West, the concern that motherhood relegates one to a life of undesired servitude and self-sacrifice is unfounded. In fact, this "version" of motherhood only became a threat for Indigenous women after contact. Indeed, according to Kim Anderson, motherhood is/was "the source of Indigenous female authority in the family and in the governance of our pre-colonial nations."[54] That is to say that the Western feminist anti-essentialist concern about the loss of power and authority through motherhood does not apply. It is/was a position and role that was desired and respected.

One can see, then, how Ruddick's notion of lucid knowledge would apply to the situation of the women water protectors as mothers. Ruddick sees lucid knowledge as a key feature of a mother's love and care for her child. As a tool of nonviolent resistance, lucid knowledge comes first in the form of knowing all one can about the person or being that is in danger or missing, then in the form of using all nonviolent means possible to gain additional knowledge. Furthermore, lucid knowledge involves publicizing the danger to others who are not aware of the situation. For the water protectors, this involves educating others about our reciprocal relationship with water. On the one hand, the protectors recognize water and Mother Earth as the givers of life—it is incumbent upon us that we care for them. On the other hand, in the care and protection of water as a life-giving entity, the women water protectors are also keenly aware of the effects of water degradation on the lives of their own children, grandchildren, and generations to come. In other words, their keen knowledge—and the sharing of this knowledge with anyone who will listen, which includes bringing non-mothers to awareness—is in fact a form of lucid knowledge to protect their environment and their loved ones at the same time. In this way, lucid knowledge and the principle of knowledge, just described, also overlap with the next concept we will discuss: interconnectedness and reciprocal responsibility.

All Our Relations

Deborah McGregor (Anishinaabe from Whitefish River First Nation) describes the ancestral and intricate relationship between Indigenous peoples and water:

> Water transcends time and space. In some respects, the waters we interact with in the present are the same waters our ancestors experienced, and the same ones that may be experienced by future generations in turn, should we take care of the waters sufficiently to ensure their (and our) future viability.[55]

Meredith Privott discusses Standing Rock in terms of its participants evoking an *ethos of responsibility*. Privott identifies three key components of this ethos: "responsive care in/to the interconnectedness of life, the special role of women in the care of water, and the collective survival of Indigenous women in colonial and patriarchal violence."[56] Although all three components are present within this conversation, the focus here will be primarily on the first two components. The notion of "responsive care" and the role of women as protectors of water is also acknowledged by other authors.[57] For example, Kyle Powys Whyte points out the impact of water's reciprocity with life, and women's particular role in protecting it, in the Anishinaabe tradition.[58] This is, in part, based upon the cosmological connection between women and water as creators of life found in many Indigenous cultures.[59]

One of the foundations of this ethos of responsibility can also be found in the cosmology of many Indigenous peoples, namely, the interconnectedness of all things. A phrase that often enters into the discourse and is well known through the work of Winona LaDuke, including her book of the same title, is *All Our Relations*. Three sets of relations frequently referred to are relations between humans and the land or environment, relations between living humans, and relations to past and future generations.

Relations to Land and the Environment

Ann Hoisington (Tingling Star Woman, Standing Rock Sioux) describes the connections to Mother Earth in an interview at Sacred Stone Camp: "As Native people, being connected with Mother Earth and the Great Spirit, we have connection with everything around us. The trees, the grass, the plants, even the rocks . . . and without water, we're nothing, because we're made of water too."[60] The connections Hoisington cites belie two ethical concerns about the pipeline. One that is of central importance to the issue of the pipeline crossing the Missouri and Cannonball Rivers is environmental contamination. At one point, Hoisington says, "it will always leak" and for that reason, we need to "fight . . . the oil, coming through our water." Unfortunately, Hoisington's view—well substantiated by past leaks—was also prescient. As previously mentioned, the Dakota Access Pipeline leaked at least five times in 2017. The biggest was a 168-gallon leak near DAPL's endpoint in Patoka, Illinois, on April 23. According to federal regulators, no wildlife was impacted, although soil was contaminated, requiring remediation.[61] There are of course reasons to believe that spills in the future could be a lot worse. The Keystone Pipeline leaked over 380,000 gallons of oil in Edinburgh, North Dakota, in November 2019.[62]

Relations to Each Other

Above and beyond the environmental consequence of the oil contaminating the water, there is, as Hoisington noted, the effect on humans who count on the Missouri as a source of water as well. Importantly, Hoisington ends with two more notes of connection that might at first glance strike us as unusually conciliatory in what is sometimes depicted as a "battle." The first has to do with recognizing the reality that, for better or worse, the need for fossil fuels will not disappear in the near-term future. But there are other humans affected by the pipeline in less obvious ways: "And if they want the oil, just don't go through the water, that's all we're asking. . . . And it takes away jobs too. It takes away jobs from the truck drivers." Even in the midst of what amounts to an environmental and spiritual crisis, Hoisington is thinking about loss of jobs a pipeline might bring. Furthermore, she even invites those pushing for the pipeline (and thus fighting against the protectors at Sacred Stone) to join her. "If they'd be willing to come down here and live, then maybe they would know. They've got us portrayed as savages."[63] But, she notes, the gunshots reported by the private security and sheriffs were in reality drums, misheard. And when there was talk overheard of "pipes" it was about *peace* pipes, not pipe *bombs* as was misreported.[64] In other words, in Hoisington's view, as well as in Privott's view and also in ours, the protectors were trying to maintain relationships, whereas many in the governor's and the Morton County sheriff's offices were depicting it otherwise.[65]

Relations to Future and Past Generations

Hoisington also connects the value of the water to the existence of future generations: "Don't put [the pipeline] through the water, because this is all we have left . . . and we want to live, I want my children to live, I want their children to live, and I want their children's children to live." Hoisington brings the multiple connections together as she points at the ground and argues, "She's still my mother, and everybody should think that way, because we all come from her, to save her and the water."[66] This last conciliatory note has the echo of centuries of Indigenous people trying to reach across worldviews toward settlers, and centuries of being misunderstood. Laura Donaldson (Cherokee) discusses the life and work of Nah-ye-hi (Cherokee), also known as Nancy Ward. Donaldson quotes Nan-ye-hi in a speech to U.S. treaty commissioners at Holston:

> You know that women are always looked upon as nothing; but we are your mothers; you are our sons. Our cry is for peace; let it continue. This peace must last forever. Let your women's sons be our sons; our sons be yours. Let your women hear our words.[67]

Donaldson instructs how to think of this first line: "You know" should be taken to mean "you," the white settler, "think that women are nothing." In the next line, "But" should be taken to signify the Cherokee perspective: "we are your mothers." The effect that Nan-ye-hi's speech had was both important and unique:

> This unparalleled act of permitting a woman to speak in the negotiating council took the commissioners aback. In their response, Colonel William Christian acknowledged the emotional effect her plea had on the men and praised her humanity, promising to respect the peace if the Cherokees likewise remained peaceful.[68]

Nan-ye-hi was a *ghigau*, or a strong female counselor, who was trying to argue against further violence on both sides. Nan-ye-hi's speech again demonstrates that the Indigenous feminist perspective cuts across three types of relations: relations to each other, relations to the land, and relations to past and future generations. Elsewhere Nan-ye-hi argues for sustaining "the land we now have, which God gave us to inhabit, and raise provisions," in other words, raise and maintain children.[69] The echo of Nan-ye-hi's call can be heard centuries later in the women water protectors of Standing Rock and elsewhere, such as Tingling Star Woman. Now that we have an understanding of the types of interconnections that the water protectors recognize, we will further analyze the notions of reciprocal responsibility in the form of systems of responsibility and collective continuance.

Reciprocal Responsibility

Deborah McGregor reminds us that "water mediates interactions among many living beings on earth. Consequently, water is considered a relative that has responsibilities to give and support life."[70] In the narratives above there exists an Indigenous environmental ethics of care. Kyle Powys Whyte and Chris Cuomo lay out five elements of this ethics of care in their chapter entitled "Environmental Ethics of Care" within the *Oxford Handbook to Environmental Ethics*.[71] Here we want to focus on the first two:

1. Emphasize the importance of awareness of one's place in a web of different connections spanning many different parties, including

humans, non-human beings and entities (e.g., wild rice, bodies of water) and collectives (e.g., forests, seasonal cycles);

2. Understand moral connections as involving relationships of interdependence that motivate reciprocal responsibilities.[72]

The first feature above corresponds to what Whyte elsewhere calls systems of responsibility. Whyte states, "*Systems of responsibilities* are the actual schemes of roles and relationships that serve as the background against which particular responsibilities stand out as meaningful and binding."[73] While Whyte goes to great lengths to acknowledge that this special connection emerges out of a specifically Anishinaabe creation story, the idea that water is life is echoed in the worldview of other Indigenous peoples as well. For instance, in Peru, where women from many Amazonian tribes are dealing with many of the same battles as the Standing Rock Lakota, Flor de María Paraná from the Indigenous Kukama community of Cuninico, Maranon River, says: "Water is life. The river is like a mother we have and water is a mother of life for human beings. Now we are demanding our rights in the face of the pollution which Petroperu has caused with crude oil, ruining our mother, natural water, which is for the consumption of all human beings."[74]

The second feature above—involving responsibility and interdependence—involves what Whyte elsewhere calls collective continuance. By collective continuance Whyte means "a community's aptitude for being adaptive in ways for the livelihoods of its members to flourish into the future." Though Whyte does not make this explicit in his definition, it is clear that the "collective" necessarily includes nonhuman community members. Elsewhere he notes, "A community may have a responsibility to care for salmon habitat; salmon, in turn, may provide food and support for other species." Put another way, Whyte notes that the "future flourishing of indigenous lives" involves connections "to the earth and its many living and nonliving beings and natural independent collectives." The important point here is that an Indigenous ethics of care is by necessity an Indigenous *environmental* ethics of care, to the extent that the flourishing of Indigenous and non-Indigenous lives alike is dependent on the collective continuance of lived environments. Sadly, the primary difference between an "ethics of care" and an Indigenous ethics of care is that most "ethics of care" are myopically focused only on human-to-human relations and miss how those relations interdepend on *all our relations*. Whyte notes that collective continuance also includes "the future flourishing of Indigenous lives that are closely connected to the earth and its many living and nonliving beings and natural interdependent collectives."[75] McGregor provides a good example of collective continuance: "We must look at the life that water supports (plants/medicines, animals, people, birds, etc.) and the life that supports water (e.g., the earth, the rain, the fish). Water has a role and responsibility to fulfill, just as people do."[76] McGregor sees water's bestowal of life-giving support to humans as necessitating a reciprocal relation. By virtue of water's responsibility to us, and all of creation, we have a responsibility to it. Obviously, it is this latter formulation that is not well understood in most non-Indigenous understandings of duties to the environment. "Collective continuance is promoted, then, both in terms of maintaining the persisting systems of responsibilities, but also in making strides toward better, responsibility-based systems of coexistence."[77]

Interconnectedness and the Maternal

As demonstrated thus far, each one of the themes being addressed are, in the worldview of most Indigenous women, deeply intertwined. In the words of Temryss MacLean Lane, "the duty of frontline warriors is to protect water, all relations, and the life that sustains our existence."[78] That is to say, the Indigenous knowledge is of *interconnected* life, one key element of which is maternal care for the land and water. Privott points out that an "ethos of responsibility can also include the distinct authority of women to care for the water . . . and [their] knowledge of water (particularly rivers) and its role in shaping life on the plains for over a thousand years."[79] Consider the centrality of water in the role of women's lives as given in the account of Nokhom Maria Campbell (Cree/Métis). Campbell notes that water is used both in birth and in "washing and preparing the dead for burial." She goes on: "Younger women have the capacity to hold life-giving waters within, and older women are responsible for helping people move through those waters in times of transition."[80] A focus on interconnectedness and reciprocal responsibility can also be seen as an important element in maternal thinking in the form of the awareness of the pain of others.

Relations and Awareness of the Pain of Others

In Ruddick's application to the Madres of Argentina, this awareness could be seen in the awareness of one mother toward the situation of all the other mothers in similarly horrific situations. But as applied to the maternal thinking of the water protectors, this awareness of the pain of others can be seen across all three types of relations outlined above. The first of these relations to the land would connect to the awareness of the pain of nonhuman others, Mother Earth, and the land. One example can be seen in the words of Jasilyn Charger (Cheyenne River Sioux Tribe), who is also founder of the Indigenous International Youth Council. Charger draws a clear connection between protecting the water and resisting violence against women: "I became involved in this movement because it is so close to my home. . . . It is something I just felt that I've been drawn to as a woman because of our connection to the earth. As a life-giver, it is something that I felt I needed to do as a woman." Charger notes "a direct connection" between violence against women and "what is happening to the earth. We face it every day, having things forced upon us in our daily lives. We know how she feels. . . . We know what it feels like to have things pushed upon us without asking, and we stand in solidarity with her, Mother Earth."[81]

A final example of this awareness can be seen in an example given by Winona LaDuke regarding pollution in the Missouri River: "My sister fished a gar out of the river, a great prehistoric fish, only to find it covered with tumors." Of course, this example is also an example of lucid knowledge: "The saying is 'The solution to pollution is dilution.' This is convenient, but not true. Blacktail Creek and Little Muddy River were contaminated after nearly three million gallons of saltwater with elevated levels of chloride contamination [entered them]. All was diluted. But then there was that gar fish with the tumors."[82] The awareness of the way specific contamination occurs—where it comes from and what it does to other beings—is a perfect example of how lucid knowledge and the awareness of the pain of others intersect naturally in an Indigenous environmental ethic.

As for the awareness of the pain of others in human relationships, we have seen multiple examples in the thinking of Ann Hoisington. Also known as Tingling Star Woman, Hoisington thought of the effects of the proposed pipeline on the unemployment of truck drivers, as well as how it might affect the water for her children and grandchildren. At Standing Rock, the two ethical principles of knowledge and interconnectedness are enacted in the form of a women's politics of resistance.

Women's Politics of Resistance and Nonviolent Indigenous Resistance

As has been noted throughout, Indigenous women have responsibilities to water. Incorporating these cultural responsibilities, there are similarities between the structure of women at Standing Rock and what Ruddick calls a "women's politics of resistance." A women's politics of resistance has three key features: it is largely organized by women (though men can be included), it uses ideas or symbols of the feminine and maternal in its approach, and it is against dominant (and unjust) power structures.[83]

Women Leaders

A women's politics of resistance is just that: a political resistance movement headed by women. Even though people of all genders and colors showed up at Standing Rock to protect the water, it was women figures such as LaDonna Brave Bull Allard, Faith Spotted Eagle (Yankton Sioux Nation), and Gracey Claymore (Hunkpapa Lakota), known as the "Women Water Warriors of DAPL," who were representatives of the movement. Brave Bull Allard established the Sacred Stone Camp in 2016. Spotted Eagle is coauthor of a statement from "the Braveheart Society, the White Buffalo Calf Woman Society and Stone Boy Society decrying the 'war of "bio-politics" being waged on indigenous homelands all across the Americas.'" Claymore met with candidate Hillary Clinton in 2016 after attending the National Congress of American Indians to discuss Clinton's help with DAPL when she was presumed to be the next president. In addition to the above-mentioned "Women Water Warriors of DAPL," a review of the interviews and press conferences available at the #NoDAPL Archive shows that women were central to representing and leading Standing Rock.[84]

Maternal Symbols

In addition to many leaders of the #NoDAPL movement being women, they also draw on the feminine and maternal to justify their resistance. Remember Ann Hoisington's call that "She [the earth] is still my Mother and everybody should think that way, because we all come from her to save her and the water."[85] But also, as we have noted, some Indigenous and First Nations peoples see women in particular as essentially connected to the protection of water. Privott describes the Walkerton Commission's inquiry and Deborah McGregor's position in it at length. In that inquiry McGregor points out that for Indigenous and First Nations people,

> water is "the giver of life with which babies are born" and as such must always be shown respect and kept clean so that "it can continue to fulfill its purpose." . . . No one is exempt from caring and paying respect to water, but for women there is a special responsibility.[86]

Perhaps the most telling indication of the way motherhood and feminine symbols play a role at Standing Rock is found in the example of Zintkala Mahpiya Wi Blackowl, Sky Bird Woman, of the Sicangu and Ihanktonwan Lakota tribes, who gave birth to "her daughter in a tipi at Oceti Sakowin, as her family slept nearby, creating a sacred space for the baby to enter this world."[87] Blackowl said, "Having babies is my act of resistance." Blackowl continued: "Our reproductive rights as Native women have been taken away from us in so many ways."[88] At the core of Standing Rock's leadership remained women, which demonstrates the power of maternal and feminine traditions for its ethical force.

Nonviolent Indigenous Resistance as a Stubborn Decision

Another component of Indigenous resistance that is necessary to address is that of the warrior. In relation to the context of women's politics of resistance, Temryss MacLean Lane describes "the word 'warrior' as a feminine embodiment of bravery, protection and empowerment. Women water protectors, standing for peace, are true modern-day warriors and defenders of all existence." As has been demonstrated, Indigenous women remain the caretakers of water. In relation to this, they hold the responsibility of water warrior. "A warrior puts on her shield of protections, her prayers, her earrings, her medicine, regalia, and her paint and stands up for justice. As a collective of Indigenous women, we assert that the frontline of refusal is anywhere we choose it to be, across colonial borders and boundaries, or over and through barriers to institutional spaces of privilege."[89]

Central to these traditions is the concept of nonviolence to address issues within Indigenous communities. One example is water walks. Water walks are "an expression of the original instructions, as relevant today as they were to our ancestors centuries ago" and are meant to connect the past, present, and future of not only the water and land but also the people who have been protecting it since the beginning. Water walks are not new movements in response to the contemporary issues surrounding water in Indigenous communities. Rather, water walks have been a component of various Indigenous cultures, particularly those around the Great Lakes prior to European contact. According to McGregor, "Water Walks were not inspired by a simple or narrow political agenda, but by respecting Anishinaabek natural law and reawakening peoples' understanding of the requirements for maintaining harmonious and reciprocal relationships among beings." Although McGregor emphasizes one particular group of Indigenous peoples, we can also understand that the concept of caretaking of the water is not unique to their region but is a global concept among Indigenous peoples.[90]

McGregor continues to note that Indigenous women are protectors of the water, the land, and the people through the "concepts of love, kindness, and generosity." In order to fully understand McGregor, it is vital to define the concept of love. There are multiple meanings for love within Indigenous culture as well as non-Indigenous culture. McGregor emphasizes that love, *zaagidowin*, is a "principle for achieving well-being."[91] It also serves as the foundation for what McGregor terms "water justice." Within the mindset of water justice, it is not simply an environmental perspective of not contaminating or altering the water and physically protecting the water. It also includes respecting the waters themselves, respecting the power of water, the vulnerability of water, the history of water, the ancestry of water. However, as we have demonstrated, for Indigenous peoples, this is not an easy task. LaDonna Brave Bull Allard, Standing Rock Sioux Tribe, reminds us that "We must remember we are part of a larger story. We are still here. We are still fighting for our lives. . . . We should not have to fight so hard to survive in our own lands."[92] Through their being nonviolent warriors, the women water protectors enact a "stubborn decision."

In her example of the Madres of Argentina, Ruddick cites as the third aspect of maternal nonviolent resistance a "stubborn decision." Ruddick acknowledges that the standard "stubborn decision" of the mother can be one based in bad faith (for example, "because I said so") and can even get in the way of an authentic child-rearing experience. "But," she goes on, "'stubborn decision' takes on a new and collective political meaning when women acting together walk out of their homes to appropriate spaces they were never meant to occupy." In her own example of the Madres of Argentina, Ruddick observes, "These stubbornly decisive Argentinian . . . women, whatever their personal timidities, publicly announce that they take responsibility for protecting the world in which they and their children must live."[93] The application to women water protectors could not be more apt. Consider several facts: like the Madres, the protectors put their bodies on the line by occupying the Sacred Stone Camp, which required them, among other things, to sacrifice by leaving their "former" lives for a time; many women (and men) faced police brutality in the form of rubber bullets, water cannons, and tear gas; the women of Sacred Stone Camp very much had in mind, as did Tingling Star Woman, their children and their children's children. In short, the women water protectors used their maternal thinking to make the stubborn decision to stand against injustice, environmental and otherwise, to make the world a better place.

Standing Rock Revisited

Temryss MacLean Lane shares her story of Standing Rock:

> Standing with my brothers and sisters, staring into the eyes of the oppressor, I realized I needed to sing. I found myself waiting for other songs and drums but none would come. So, I humbly sang the only prayer song I knew. . . . And I will sing again and again and again on the frontline, however and wherever I respectfully choose, as I know it to be my duty to protect the water, honor our ancestors, and carry their teachings for the next generations yet to come. We are Indigenous women warriors, defenders of water, and protectors of life."[94]

Almost everything happening at Standing Rock was considered an act of resistance—even the nonviolent peaceful encampment. Resistance to what? Ostensibly, to Energy Transfer Partners, the company responsible for the Dakota Access Pipeline. But as both Zintkala Mahpiya Wi Blackowl and Winona LaDuke have shown us, this resistance only makes sense when put into a larger—longer—timeline. This longer timeline shows not only the injustice done to Indigenous people by the federal and state governments, but also the cozy relationship between non-Indigenous governments and corporate interests.

Consider the fact that "In mid-August [2016], Standing Rock Tribal Chairman Dave Archambault II was arrested by state police, along with 27 others, for opposing the Dakota Access Pipeline. In the meantime, North Dakota Gov. Jack Dalrymple called for more police support." In Archambault's own words: "Perhaps only in North Dakota, where oil tycoons wine and dine elected officials, and where the governor, Jack Dalrymple, serves as an adviser to the Trump campaign, would state and county governments act as the armed enforcement for corporate interests."[95]

The resistance is not only for "the dignity and future of a nation." Indeed, the Dakota Access Pipeline could have devastating effects on the environment. Already the Missouri is the "seventh most polluted river," due to agricultural runoff, fracking, and in January 2015, saltwater contamination from a pipeline spill.[96] As we pointed out above, the notion of collective continuance applies here. That is, there is an acknowledged interdependence and reciprocal responsibility between bodies of water and the people who use them, as well as the life they create and sustain. Thus, in the same vein that Nan-ye-hi once appealed to everyone who would listen about the giving away of land, Archambault urges:

> I am here to advise anyone that will listen that the Dakota Access Pipeline is harmful. It will not be just harmful to my people, but its intent and construction will harm the water in the Missouri River, which is one of the cleanest and safest river tributaries left in the United States. To poison the water is to poison the substance of life. Everything that moves must have water. How can we talk about and knowingly poison water?[97]

Importantly, not once does Archambault say "my" or "our" water, reinforcing the view that water is its own form of life and has its own being. It should only be on the basis of symbiosis that we could talk about "using" water. "Facing militarized violence, Standing Rock's leadership put out a call-to-action that sparked a growing movement of solidarity for Native peoples to stand together as human beings and water protectors."[98] The call went far and wide as Indigenous and non-Indigenous peoples came together to fight the water injustice taking place at Mni Sose. Standing together at Standing Rock is not where water protection ends. The movement of water protection among Indigenous peoples extends throughout North America and extends to non-Indigenous people as well.

Becoming Water Protectors

The water protectors demonstrate an Indigenous women's politics of resistance as part of an ongoing history. In the words of McGregor, "People must relate to water in order to live. . . . All humanity shares this basic need for survival: at a fundamental level we need water to live."[99] Not only does the Sacred Stone camp exemplify an Indigenous women's environmental ethics of care, but it also serves as a model for future Indigenous and non-Indigenous environmental movements.

> Despite lacking participation in policy processes, some Anishinaabe women are taking collective action to carry out their responsibilities to water. As previously referred to, McGregor discusses how a group of Anishinaabe women began walking around the Great Lakes in the early 2000s, which they call the Mother Earth Water Walks. The purpose is to help people in the basin recognize and re-recognize the importance of water in its spiritual dimensions instead of as an inanimate resource.[100]

The Mother Earth Water Walks are not meant just for Indigenous peoples, but rather embody a movement for all peoples who are looking to respect, reciprocate, and continue collectively with the waters. This responsibility, particularly attributable to the youth, was fully represented at Standing Rock.

However, this responsibility does not end, nor did it even begin, with Standing Rock. Standing Rock simply demonstrated for non-Indigenous people that showing up as allies was only the beginning. We must all, as humanity, participate in sharing the responsibility for protecting water. We all can be water protectors through the guidance of those who came before us and are now establishing organizations and means for assisting. One such organization is Honor the Earth, an environmental organization founded in 1993 by Winona LaDuke and Indigo Girls Amy Ray and Emily Saliers. LaDuke is well known throughout Indigenous and environmental communities as a leader in sustainability. Honor the Earth states that their mission "is to create awareness and support for Native environmental issues and to develop needed financial and political resources for the survival of sustainable Native communities." As an Indigenous organization, their mission and goals are deeply rooted in their "communities, histories, and long-term struggles to protect the Earth."[101] However, they go beyond Indigenous issues as well in seeking to create a network of individuals, Indigenous and non-Indigenous, who, like themselves, value the environment beyond a physical entity.

Recently, Honor the Earth implemented a new effort entitled the Water Protector Education and Culture Project:

> Through this project, we are creating spaces for Water Protectors to gather and learn about our lands, our traditions, and models for a new sustainable future. We are creating the space for our people to practice our way of life on our land, and lead traditional cultural teachings.[102]

Although based within an Indigenous organization and initiative, the program is open for ALL peoples to join. The principles of *All Our Relations* do not end with Indigenous peoples, it includes ALL peoples. And therefore, all peoples can become water protectors. There are multiple ways and reasons for protecting the waters around us as "our way of life." "The water, plants, and animals here are under attack by a system that sees the earth as a source of profit and not our mother to respect and care for."[103] Therefore, not only must we all become water protectors, it is our responsibility to become water protectors due to our reciprocal relationship to water. The water and earth care for us only as long as we walk the correct path. Honor the Earth reminds us that we are in what the Anishinaabe prophecies call "the time of the Seventh Fire. This is a time when our people will have two roads ahead of us—one *miikina*, or path, which is well-worn—but scorches—and another path which is green. It will be our choice upon which path to embark. That is where we are."[104]

Once we realize that choice—between the scorched path and the green one—we realize that we must ALL take responsibility for the waters around us, and do so under the guidance of Indigenous knowledge and care. However, for those who are non-Indigenous, one must adhere to the fact that one is "stepping into an area where you are a guest" and therefore be willing to understand and be respectful of the cultures one is entering into and conduct oneself appropriately.[105]

Being an ally and water protector not only means standing beside Indigenous peoples; it means respecting their cultures, learning their histories, and recognizing one's privilege when entering their communities and lands. This encompasses understanding the community values in relation to the environment. Although specific from Honor the Earth and based in Ojibwe cultures, the seven values below are found among most Indigenous communities and should be adhered to as one becomes a water protector:

- Honesty—To achieve honesty within yourself is to recognize who and what you are. Do this and you can be honest with all others.
- Humility—Humble yourself and recognize that no matter how much you think you know, you know very little.
- Truth—To learn Truth, to live Truth, to walk Truth, and to speak Truth.
- Wisdom—To have Wisdom to know the difference between good and bad and to know the result of your actions.
- Love/Compassion—Unconditional love and compassion is to know that when people are weak, they need your love and compassion the most.
- Respect—Respect others, their beliefs. Respect yourself; when you practice respect, respect will be given back to you.
- Bravery/Courage—Have bravery and courage in doing things right even though it may hurt you physically and mentally.

In addition, it is also noted that to be a good water protector, one must also "be a good relative" and "walk the walk." The seven values and recommendations by the Water Protectors—Honor the Earth correlate with McGregor stressing that "humanity is the recipient of the generosity of the Earth, and in turn, we must share and give something of ourselves back to the Earth, including the water."[106]

Conclusion

All the resistance of the water protectors at Standing Rock—by women and men alike—was not in vain. At first, there was hope that the project would not be completed, when the Obama administration decided against issuing permits in December 2016. Unfortunately, the Trump administration reversed this decision after only two days in office, signaling what has become a very cozy relationship with fossil fuel companies. Construction of the pipeline itself was completed in June 2017, but the Standing Rock Sioux did not stand down. They took up the issue of not having a proper say in the permitting process and continued to push through the federal legal system to bring action to stop the pipeline's operations. The relentless efforts of Indigenous people resulted in success for halting the pipeline's operations. As of this writing, numerous decisions have been made with regard to the pipeline, including a D.C. Circuit Court ruling that the pipeline be shut down. The court ordered that the pipeline be shut down while a full and proper environmental review takes place. In June 2020, the Standing Rock Sioux filed an additional brief arguing that Dakota Access should be shut down permanently, citing both the possibility of environmental damage and the lag in oil consumption more generally, in part due to the COVID-19 pandemic. On July 6, the court found in favor of the Sioux. However, the Dallas billionaire who owns Energy Transfer appealed the decision. On the original date of the shutdown, August 5, 2020, the U.S. Court of Appeals in D.C. issued an order that allowed the pipeline's operations to continue; however, they also negated the permit for the pipeline to cross Lake Oahe. Essentially, the ruling allowed the pipeline to continue functioning, but noted at the same time its illegality. The Standing Rock people did not accept this ruling and continue to push for a full shutdown of the pipeline. As of the time of this writing, the U.S. Army Corps of Engineers has been tasked to complete a full environmental review for the permits to be legitimately reissued. "A final decision on whether to re-issue permits for the Dakota Access Pipeline is unlikely until after the 2020 presidential election, which sets up the possibility of a permanent closure."[107]

The work the water protectors did is not inspirational simply because it appears to have had some influence on favorable court decisions. Rather, the women water warriors at Standing Rock have inspired a generation of anti-fossil-fuel and anti-corporate activism. In this chapter we outlined that ethic. We began by noting the key features of an Indigenous view of water. Understanding the idea that water is life-giving, life-sustaining, and in a spiritual and poetic way, alive itself, is central to the water warriors' view and motivation. We did see, though, that the Western environmental views of bio- and eco-centrism, while not going so far as to suggest that water is alive, can ground a view of water having equal standing as a living thing, since so many living beings depend on it. We then discussed the environmental ethics of women water warriors, which included the following features. First, such an ethics involves a comprehensive knowledge of the environmental impact of human decisions, one that involves not only TEK, but historical and ecological knowledge as well. Secondly, this ethics favors a view of people and the environment as reciprocally interconnected and grounded in a mission of collective continuance that places human beings in systems of responsibility. Thirdly, we saw that this ethics is driven by nonviolent resistance, as shown in the ideal of a "warrior" of peace and well-being. Throughout, we also acknowledged the connection between these features and the centrality of motherhood to many of the Indigenous women water warriors. We compared this to Sara Ruddick's attempt for a universal "women's politics of resistance" guided by the principles of lucid knowledge, an awareness of the pain of others, and a stubborn decision to resist more powerful forces. These clearly fit with the main themes of knowledge, interconnectedness, and nonviolent resistance outlined above. Lastly, we attempted to make the case that any and all of us can be water protectors—and if we are moved by the same notions of reciprocal responsibility described above, we should be.

NOTES

1. Temryss MacLean Lane, "The Frontline of Refusal: Indigenous Women Warriors of Standing Rock," *International Journal of Qualitative Studies in Education* 31, no. 3 (2019): 197–214 at 197.

2. Lane, "The Frontline of Refusal," 197.

3. It should be noted that increasingly the ecological worldview is incorporating Indigenous knowledges and land management as a key strategy in reversing biodiversity loss. See, for example, https://www.campaignfornature.org/indigenous-peoples.

4. Nick Estes, "Fighting for our Lives: #NoDAPL in Historical Context," *Wicazo Sa Review* 32, no. 2 (2017): 115–122, at 115.

5. Lane, "The Frontline of Refusal," 201.

6. Lane, "The Frontline of Refusal," 201.

7. Alleen Brown, "Five Spills, Six Months in Operation: Dakota Access Track Record Highlights Unavoidable Reality—Pipelines Leak," *The Intercept*, January 9, 2018, https://theintercept.com/2018/01/09/dakota-access-pipeline-leak-energy-transfer-partners/.

8. Jaskarin Dhillon and Nick Estes, "Introduction: Standing Rock, #NoDAPL, and Mni Wiconi," *Fieldsights*, December 22, 2016, https://culanth.org/fieldsights/introduction-standing-rock-no-dapl-and-mni-wiconi.

9. Meredith Privott, "An Ethos of Responsibility and Indigenous Women Water Protectors in the #NoDAPL Movement," *American Indian Quarterly* 43, no. 1 (Winter 2019): 74–100.

10. Winona LaDuke, "What Would Sitting Bull Do?" *Yes! Magazine*, August 29, 2016, https://www.yesmagazine.org/environment/2016/08/29/an-oil-pipeline-and-a-river-what-would-sitting-bull-do/.

11. Audrea Lim, "'The Next Standing Rock': Minnesota's Indigenous Water Protectors Are Already Camping to Defeat Line 3," *The Progressive* (December 2017/January 2018): 18–20, at 19.

12. Lane, "The Frontline of Refusal," 197.

13. As quoted in Kate Cave and Shianne McKay, "Water Song: Indigenous Women and Water," *Solutions* 7, no. 6 (November 2016): 64–73, https://www.thesolutionsjournal.com/article/water-song-indigenous-women-water/.

14. Tiokasin Ghosthorse, "Living in Relativity," Center for Humans & Nature, August 26, 2016, https://www.humansandnature.org/living-in-relativity.

15. Ghosthorse, "Living in Relativity."

16. Edward Benton-Banai, as quoted in Cave and McKay, "Water Song."

17. Karletta Chief, Alison Meadow, and Kyle Whyte, "Engaging Southwestern Tribes in Sustainable Water Resources Topics and Management," *Water* 8, no. 8 (August 2016): 350.

18. "Life: Life Definition," Biology Online, www.biologyonline.com/dictionary/life.

19. Molly Sargen, "Biological Roles of Water: Why Is Water Necessary for Life?," *Science in the News* (blog), September 26, 2019, https://sitn.hms.harvard.edu/uncategorized/2019/biological-roles-of-water-why-is-water-necessary-for-life/.

20. Kenneth Goodpaster, "On Being Morally Considerable," in *Environmental Ethics: Readings in Theory and Application*, 5th ed., ed. Louis Pojman and Paul Pojman (Belmont, CA: Wadsworth, 2006), 154–163.

21. Goodpaster, "On Being Morally Considerable," at 161.

22. Aldo Leopold, "The Land Ethic," in Pojman and Pojman, *Environmental Ethics*, 163–172, at 172.

23. J. Baird Callicott, "Animal Liberation: A Triangular Affair," *Environmental Ethics* 2 (Winter 1980): 311–338, at 323, emphasis added.

24. Callicott, "Animal Liberation," 324.

25. J. Baird Callicott, "The Conceptual Foundations of the Land Ethic," in Pojman and Pojman, *Environmental Ethics*, 173–185, at 182.

26. It's also worth noting another study done by Callicott, with Michael P. Nelson, *American*

Indian Environmental Ethics (Upper Saddle River, NJ: Prentice Hall, 2004). After first relaying Ojibwe narratives that concern the origins and treatment of nature, Callicott and Nelson spend several pages of their interpretive essay (121–132) discussing the possible relations between Ojibwe environmental views and the land ethic. We haven't included that only because they do not have an explicit discussion of water there, outside of its general inclusion in biotic communities.

27. Leopold, "The Land Ethic," 170, 169.

28. Robin Wall Kimmerer, *Braiding Sweetgrass: Indigenous Wisdom, Scientific Knowledge, and the Teachings of Plants* (Minneapolis: Milkweed Editions, 2013), 9.

29. J. T. Inglis, ed., *Traditional Ecological Knowledge: Concepts and Cases* (Ottawa: Canadian Museum of Nature, 1993), vi.

30. Vine Deloria Jr., "Ethnoscience and Indian Realities," in *Spirit and Reason: The Vine Deloria, Jr. Reader*, ed. B. Deloria, K. Foehner, and S. Scinta (Golden, CO: Fulcrum Publishing, 1999), 63–71, at 63.

31. F. Berkes, "Traditional Ecological Knowledge in Perspective," in *Traditional Ecological Knowledge: Concepts and Cases*, ed. J. T. Inglis (Ottawa: Canadian Museum of Nature, 1993), 1–9.

32. Kyle Powys Whyte, "The Dakota Access Pipeline, Environmental Injustice, and U.S. Colonialism," *Red Ink* 19, no. 1 (2017): 154–169, at 156.

33. See, for example, Linda Robyn, "Indigenous Knowledge and Technology: Creating Environmental Justice in the 21st Century," in *Environmental Ethics*, 7th ed., ed. Paul Pojman, Louis Pojman, and Katie McShane (Boston: Cengage Learning, 2017), 367–382. See also Kyle Powys Whyte, "Indigenous Women, Climate Change Impacts, and Collective Action," *Hypatia* 29, no. 3 (2014): 599–616.

34. Lynn White, "The Historical Roots of Our Ecological Crisis," in Pojman, Pojman, and McShane, *Environmental Ethics*, 5th ed., 14–21.

35. Kathryn Akipa (Sisseton-Wahpeton Oyate) et al., "Reflections on Mni Sose after Lewis and Clark," in *This Stretch of the River*, ed. Craig Howe and Kim TallBear (Sioux Falls, SD: Pine Hill Press, 2006), as quoted in Privott, "An Ethos of Responsibility," 84.

36. LaDuke, "What Would Sitting Bull Do?"

37. Amy Dalrympyle, "Pipeline Route Plan First Called for Crossing North of Bismarck," *Bismarck Tribune*, August 18, 2016.

38. T. J. Raphael, "Bismarck Residents Got the Dakota Access Pipeline Removed without a Fight," *The Takeaway*, December 1, 2016, https://www.pri.org/stories/2016-12-01/bismarck-residents-got-dakota-access-pipeline-moved-without-fight.

39. Catherine Thorbecke, "Why a Previously Proposed Route of Dakota Access Pipeline

Was Rejected," *ABC News*, November 3, 2016, https://abcnews.go.com/US/previously-proposed-route-dakota-access-pipeline-rejected/story?id=43274356.

40. "Treaties Still Matter: The Dakota Access Pipeline," *Native Knowledge 360*, https://americanindian.si.edu/nk360/plains-treaties/dapl.

41. Winona LaDuke, *All Our Relations: Native Struggles for Land and Life* (Cambridge, MA: South End Press, 1999), 17.

42. LaDuke, *All Our Relations*, 18–19.

43. Sara Ruddick, *Maternal Thinking: Toward a Politics of Peace* (Boston: Beacon Press, 1995), 228.

44. See Ruddick, *Maternal Thinking*, 230. Ruddick gets these characteristics from Phillip Hallie writing about a group of French parishioners who resisted the Nazis nonviolently: "Lucid knowledge, awareness of the pain of others, and stubborn decision dissipated for the Chambonnais the Night and Fog that inhabited the minds of so many people in Europe, and the world at large, in 1942."

45. Lester Kurtz, "The Mothers of the Disappeared: Challenging the Junta in Argentina (1977–1983)," *International Center for Nonviolent Conflict*, July 2010, https://www.nonviolent-conflict.org/wp-content/uploads/2016/02/The-Mothers-of-the-Disappeared-Argentina-7.pdf.

46. Kurtz, "The Mothers of the Disappeared," 350.

47. Ruddick, *Maternal Thinking*, 230.

48. Josh Fox et al., International WOW Company, Digital Smoke Signals, Morninglight Films, and Bullfrog Films, *Awake: A Dream from Standing Rock* (Oley, PA, 2017).

49. See the introduction (and *passim*) to Simone de Beauvoir, *The Second Sex*, trans. H. M. Parshley (New York: Vintage, 1989). The potential for this type of criticism arises out of the second-wave feminist Simone de Beauvoir's *The Second Sex*, in which Beauvoir argues that to conceive of women as mothers only is to reduce their myriad possibilities and thus limit their freedom.

50. For an excellent summary of the debate surrounding the accusation that Ruddick's "Maternal Thinking" is ethnocentric, see Jean Keller, "Rethinking Ruddick and the Ethnocentrism Critique of 'Maternal Thinking,'" *Hypatia* 25, no. 4 (2010): 834–851. In that work Keller identifies two primary sources of criticism: Maria Lugones and Alison Bailey. See Maria Lugones, "On the Logic of Pluralist Feminism," in *Pilgrimages/Peregrinajes: Theorizing Coalition against Multiple Oppressions* (New York: Rowman & Littlefield, 2003); Alison Bailey, "Mothering, Diversity, and Peace Politics," *Hypatia* 9, no. 2 (1994): 188–198. Lugones argues that identifying one's social position cannot be a substitute for actually interacting with and involving the views of other cultures. Bailey

argues "against universal goals [for mothering] by demonstrating that within the United States, racial ethnic maternal practice is guided by its own separate and distinctive set of constitutive goals" (193). Keller ultimately acknowledges that while Ruddick's account of mothering practices is necessarily incomplete, it was never Ruddick's aim to be exhaustive in her description of mothering practices, and instead it was her intention to attempt to find the commonalities among them (846). Her answer is a "modified universalism," which ultimately finds some fault with Ruddick's limited goals—all child-centric—but nonetheless acknowledges that some specific goals of mothering can be universalized (847).

51. Ruddick, *Maternal Thinking*, xii.

52. Indeed, this may not be sufficient to quash concerns of ethnocentrism, to the extent that one may not adequately allow the recognition of the practices of other cultures to influence one's own account. Still, as we are suggesting with reference to Keller's "Rethinking Ruddick" above, the accusation of Ruddick beingly unredeemably ethnocentric doesn't hold up under scrutiny.

53. Verna St. Denis, "Feminism Is for Everybody: Aboriginal Women, Feminism and Diversity," in *Making Space for Indigenous Feminism*, ed. Joyce Green (Black Point, NS: Fernwood Publishing, 2007), 38.

54. Kim Anderson, "Affirmations of an Indigenous Feminist," in *Indigenous Women and Feminism*, ed. Cheryl Suzack et al. (Vancouver: UBC Press, 2010), 86.

55. Deborah McGregor, "Indigenous Women, Water Justice, and *Zaagidowin* (Love)," *Canadian Woman Studies/Les Cahiers de la Femme* 30, nos. 2 and 3 (2013): 71–78, at 73.

56. Privott, "An Ethos of Responsibility," 74.

57. Deborah McGregor, "Honouring our Relations: An Anishinaabe Perspective on Environmental Justice," in *Speaking for Ourselves: Environmental Justice in Canada*, ed. J. Argyeman, P. Cole, and R. Haluza-Delay (Vancouver: University of British Columbia Press, 2009); see also Whyte, "Indigenous Women."

58. Whyte, "Indigenous Women."

59. Indigenous women and scholars have no more interest in being ethnocentricists than any right-thinking person would. In other words, most Indigenous women acknowledge that even though they may believe that womanhood, motherhood, creation, and water protection are interwoven concepts, they have no right to impose that view on others, nor assume it. That said, we think it is accurate to say that as *women* most Indigenous water protectors would argue that *water is life* on the basis of those beliefs, and thus other non-Indigenous people ought to recognize that as well. In other words, the same conclusion, that water is life, can be drawn for different reasons—and with different

supports—as we try to show elsewhere in this chapter.

60. Privott, "An Ethos of Responsibility," 81.

61. Brown, "Five Spills."

62. Morgan Winsor, "Over 380,000 Gallons of Oil Spill from Keystone Pipeline in North Dakota," *ABC News*, November 1, 2019, https://abcnews.go.com/US/380000-gallons-oil-spill-keystone-pipeline-north-dakota/story?id=66683075.

63. Privott, "An Ethos of Responsibility," 83.

64. Privott, "An Ethos of Responsibility," 83.

65. "Media Manipulation: Narratives Perpetuating Negative Stereotypes of Native Americans, Water Protectors, and the #NoDAPL Movement in General," #NoDAPL Archive, https://www.nodaplarchive.com/media-manipulation.html.

66. "Media Manipulation."

67. Laura Donaldson, "'But We Are Your Mothers, You Are Our Sons': Gender, Sovereignty, and the Nation in Early Cherokee Women's Writing," in *Indigenous Women and Feminism*, ed. Cheryl Suzack et al. (Vancouver: UBC Press, 2010), 43–55.

68. Continued: "In July 1781 she spoke powerfully at the negotiations held on the Long Island of the Holston River following settler attacks on Cherokee towns. Oconastota designated Kaiyah-tahee (Old Tassel) to represent the Council of Chiefs in the meeting with John Sevier and the other treaty commissioners. After Old Tassel finished his persuasive talk, Ward called for a lasting peace on behalf of both white and Indian women." David Ray Smith, "Nancy Ward," *Tennessee Encyclopedia*, updated March 1, 2018, https://tennesseeencyclopedia.net/entries/nancy-ward/.

69. Donaldson, "'But We Are Your Mothers,'" 47.

70. As quoted in Whyte, "Indigenous Women," 605.

71. Kyle Powys Whyte and Chris Cuomo, "Ethics of Caring: Indigenous and Feminist Philosophies," in *The Oxford Handbook of Environmental Ethics*, ed. Stephen M. Gardiner and Allen Thompson (New York: Oxford University Press, 2017), 235–247.

72. Whyte and Cuomo, "Ethics of Caring," 236.

73. Whyte, "Indigenous Women," 603.

74. Sophia Pincetti, "'The River Is a Mother': Four Indigenous Women Water Protectors from Peru's Amazon," *Chaikuni Institute*, March 8, 2018, https://chaikuni.org/news/the-river-is-a-mother-4-indigenous-women-water-protectors-from-peru-s-amazon.

75. Whyte, "Indigenous Women," 602, 603.

76. McGregor, "Honouring our Relations," 37–38, as quoted in Whyte, "Indigenous Women," 605.

77. Whyte, "Indigenous Women," 607.

78. Lane, "The Frontline of Refusal," 199.

79. Privott, "An Ethos of Responsibility," 83.

80. Kim Anderson, Barbara Clow, and Margaret Haworth-Brockman, "Carriers of Water: Aboriginal Women's Experiences, Relationships, and Reflections," *Journal of Cleaner Production* 60, no. 1 (2013): 11–17, at 13.

81. Privott, "An Ethos of Responsibility," 90.

82. LaDuke, "What Would Sitting Bull Do?"

83. Ruddick, *Maternal Thinking*, 223.

84. "#NoDAPL Archive—Standing Rock Water Protectors," #NoDAPL Archive, https://www.nodaplarchive.com.

85. Privott, "An Ethos of Responsibility," 83.

86. From McGregor, "Honouring our Relations," as quoted in Privott, "An Ethos of Responsibility," 87.

87. "Women Warrior Water Protectors of DAPL," *Indian Country Today*, March 8, 2017, https://indiancountrytoday.com/archive/women-water-protectors-dapl.

88. "Women Warrior Water Protectors of DAPL."

89. Lane, "The Frontline of Refusal," 197, 212.

90. McGregor, "Indigenous Women," 74.

91. McGregor, "Indigenous Women," 71, 72.

92. Whyte, "The Dakota Access Pipeline," 154.

93. Ruddick, *Maternal Thinking*, 233.

94. Lane, "The Frontline of Refusal," 214.

95. LaDuke, "What Would Sitting Bull Do?"

96. LaDuke, "What Would Sitting Bull Do?"

97. LaDuke, "What Would Sitting Bull Do?"

98. Lane, "The Frontline of Refusal," 199.

99. McGregor, "Indigenous Women," 71.

100. Whyte, "Indigenous Women," 606.

101. "About Us," Honor the Earth, http://www.honorearth.org/about.

102. "Welcome Water Protectors," Honor the Earth, http://www.honorearth.org/welcome_water_protectors.

103. "Welcome Water Protectors."

104. "About Us."

105. "How to Be a Water Protector," Water Protectors–Honor the Earth, https://welcomewaterprotectors.squarespace.com/respectful-water-protector.

106. McGregor, "Indigenous Women," 76.

107. Alyssa Schukar, "D.C. Circuit Issues Mixed Decision on Dakota Access Shutdown Order," *Earth Justice*, August 5, 2020, https://earthjustice.org/news/press/2020/dc-circuit-mixed-decision-dakota-access-shutdown-order.

SUGGESTED READINGS

Argyeman, P. Cole, and R. Haluza-Delay, eds. *Speaking for Ourselves: Environmental Justice in Canada*. Vancouver: University of British Columbia Press, 2009.

Benton-Banai, E. *The Mishomis Book: The Voice of the Ojibway*. Hayward, WI: Indian Country Communications, 1998.

Chiblow, Susan (Ogamauh anng qwe). "Anishinabek Women's Nibi Giikendaaswin (Water Knowledge)." *Water* 11, no. 2 (Winter 2019). https://doi.org/10.3390/w11020209.

Deloria, B., K. Foehner, and S. Scinta, eds. *Spirit and Reason: The Vine Deloria, Jr. Reader*. Golden, CO: Fulcrum Publishing, 1999.

Green, Joyce, ed. *Making Space for Indigenous Feminism*. Black Point, NS: Fernwood Publishing, 2007.

Howe, Craig, and Kim TallBear, eds. *This Stretch of the River*. Sioux Falls, SD: Oak Lake Writers Society & Pine Hill Press, 2006.

Lavalley, Giselle. *Aboriginal Traditional Knowledge and Source Water Protection: First Nations' Views on Taking Care of Water*. Toronto: Chiefs of Ontario and Environment Canada, 2006.

Nicholas, George. "An Uneasy Alliance: Indigenous Traditional Knowledge Enriches Science." *The Conversation*, February 27, 2019. https://theconversation.com/an-uneasy-alliance-indigenous-traditional-knowledge-enriches-science-109212.

Pojman, Louis, Paul Pojman, and Katie McShane, eds. *Environmental Ethics: Readings in Theory and Application*. 7th ed. Boston: Cengage Learning, 2017.

Smith, David Ray. "Nancy Ward." *Tennessee Encyclopedia*, updated March 1, 2018. https://tennesseeencyclopedia.net/entries/nancy-ward/.

Suzack, Cheryl, et al., eds. *Indigenous Women and Feminism*. Vancouver: University of British Columbia Press, 2010.

Whyte, Kyle Powys. "Indigenous Women, Climate Change Impacts, and Collective Action." *Hypatia* 29, no. 3 (2014): 599–616.

Zhao, Lucas. "Standing Rock Sioux Prevails as Federal Judge Strikes Down DAPL Permits." *Earth Justice*, March 25, 2020. https://earthjustice.org/news/press/2020/standing-rock-sioux-tribe-prevails-as-federal-judge-strikes-down-dapl-permits.

Wilbur on Drugs

Antimicrobial Use in Hog CAFOs

J. M. Dieterle and Wade Tornquist

Charlotte's *Web* is the story of Wilbur, a pig on a Midwestern farm. Wilbur is spared from slaughter by the machinations of his spider friend Charlotte and is able to live out the rest of his natural life on the farm. E. B. White ends the story with a description of Wilbur's home:

> Life in the barn was very good—night and day, winter and summer, spring and fall, dull days and bright days. It was the best place to be, thought Wilbur, this warm delicious cellar, with the garrulous geese, the heat of the sun, the passage of swallows, the nearness of rats, the sameness of sheep, the love of spiders, the smell of manure, and the glory of everything.[1]

The Midwestern United States is home to roughly 28,000 hog farms.[2] In September 2019, there were 77.7 million hogs in U.S. commercial agriculture; 56.9 million of those were in the Midwest.[3] The overwhelming majority of hogs raised in the Midwest do not have it as good as Wilbur. They don't have a warm cellar in which to sleep, they cannot feel the heat of the sun, and they cannot see the passage of swallows. Instead, they are housed indoors, in concentrated animal feeding operations (CAFOs). When they go to slaughter, pigs weigh roughly 100 kg (220 pounds) and are about 130 cm long (51 inches) and 32 cm (13 inches) wide at the shoulder.[4] Growing pigs are typically raised in community pens of ten or more animals and each is allotted, on average, 0.67 square meter (7.2 square feet) of floor space. This area is only about 60 percent larger than the rectangular dimensions of the typical pig itself (0.42 square meter) and is midway between the "most efficient economic space allocation" of 0.55 square meter and the "optimal space based on growth performance" of approximately 0.81 square meter.[5] Note that both of these two limiting measures are economic indicators. They are related to pig welfare only in a thin sense of welfare. The farmer's goal is to maximize growth performance and the price pigs fetch at market. Unhealthy pigs undermine those goals.

Housing so many pigs in such close quarters has consequences that reach beyond pig welfare. Local communities are often negatively impacted. Fulton County, Illinois, is home to at least five hog CAFOs. In 2016, Professional Swine Management LLC submitted an application to construct an additional 20,000-head hog CAFO in the county: Runway Ridge.[6] Although Illinois law gives local communities no effective power over the construction and siting of animal confinement systems,[7] Fulton County residents organized against the proposed CAFO and submitted petitions to their county commissioners. They were concerned about the odor that emanates from hog confinement systems, the pollution associated with CAFOs generally, and the changing of their rural landscape. As the *Chicago Tribune* puts it, they were fighting for their "creeks, clean air, one-lane roads, and rural way of life."[8] In this particular case, the residents had ammunition: the application had several errors (for example, the application failed to include two forks in a nearby watershed and it mislabeled several plots of land as nonresidential). In January 2017, Professional Swine withdrew its application to construct Runway Ridge. This was a rare victory for residents in the vicinity of proposed CAFOs. Sonja Eayrs, a lawyer engaged in a similar fight against the construction of CAFOs in Dodge County, Minnesota, says, "Local farm families are helpless. There is still no pathway for adversely impacted citizens."[9]

Another consequence of housing pigs indoors in close quarters is the threat of disease; in fact, hogs are highly susceptible to disease. Since large-scale confinement systems are prime locations for spreading disease, pigs were among the last farmed animals to become part of the industrial animal production system.[10] Pharmaceuticals were the key to bringing pigs inside; they allowed the industry to successfully breed and grow lots of animals in comparatively little space. Antimicrobials are particularly important enablers, because they can prevent disease from occurring and/or spreading by attenuating or completely eliminating microbe populations that live and grow within the treated animals.

Antimicrobials are used in CAFOs in three different ways: in *therapeutic* treatment of only diagnosed diseased animals, as a *metaphylactic* treatment of the herd after at least one animal has been diagnosed with a disease, and as a *prophylactic* treatment of the herd as a preventative to disease when there is no evidence of diseased animals. Prophylactic treatment is a nontherapeutic measure. Metaphylactic treatment is both a therapeutic measure for the sick animals and a nontherapeutic measure for the animals that share water and feed receptacles with the sick animals.[11]

Growth promotion is another nontherapeutic agribusiness production benefit derived from the use of antimicrobials on pigs. In the recent past, the agricultural industry routinely added antimicrobials to animal feed or water at subtherapeutic levels (typically in the range of 0.25–12.5 grams of antimicrobial per 100 kg of feed) to promote growth and enhance feed efficiency.[12] However, effective January 1, 2017, the FDA issued a rule that banned any use of antimicrobials for "production purposes," such as growth promotion.[13] But farmers may still administer antimicrobials prophylactically to "treat, control, or prevent disease."[14] And so they continue to derive the same growth promotion and production benefits as before.[15] The FDA rule has reduced the amount of antimicrobials used in agriculture generally, but the use of medically important antimicrobials to prevent disease in hog farms is still very high.[16] This is because hogs live longer than other farmed animals (six months vs. six weeks for chickens). The longer an animal lives, the more likely it is to get sick, especially when it is living in a crowded environment.[17] Globally, antimicrobial use in agriculture is rising; Van Boeckel et al. predict a 67 percent increase over 2010 levels by 2030.[18] The predicted increase is predicated on increased consumer demand for animal products and a shift to industrial farming methods in middle-income countries.

As noted above, the Midwest is home to the majority of U.S. farmed hogs. Most Midwest farms depend on antimicrobials for pig production. In this chapter, we argue that the use of antimicrobials in commercial animal agriculture should be heavily regulated, and the prophylactic (nontherapeutic) use of antimicrobials should be banned. Farmed animals should be given antimicrobials only for therapeutic purposes—that is, only when they are sick or lame. In this chapter we explain why antimicrobials are currently used in pig production and discuss the consequences of such use for human and animal health. Our primary argument begins by examining Jonny Anomaly's defense of regulating antimicrobial use on the farm.[19] Anomaly uses the classical liberal Harm Principle to defend regulation. We argue that the Harm Principle puts the burden of proof in the wrong place and suggest a shift in the way we should think about the good of antimicrobial effectiveness: as a good we share in common. The overuse of antimicrobials, we argue, is a problem of the commons. Regulation is likely to necessitate substantial changes in the way Midwest farmers raise their pigs. There are also additional likely consequences of strict regulation of antimicrobial use in commercial animal agriculture.

Down on the Hog Farm

Confinement systems offer hog farmers many benefits. Raising hogs inside protects them from both the elements and predation. The buildings allow for climate control, which enables producers to farrow the sows at any time of the year.[20] It is possible for healthy sows kept indoors to produce 2.7 litters per year. In the United States, the average number of litters per year is 2.27, with some sows giving birth to more than ten piglets per litter.[21] A further benefit for producers is that confinement systems allow them to substantially increase the number of hogs raised per farm. But by housing pigs in high population densities, farmers also increase the likelihood of spreading infectious diseases throughout a herd.

Infectious diseases are caused by the invasion of a host body by microbes. Microbes are single-celled organisms, such as bacteria, fungi, and parasites, that are found in the soil and water, in and on animals and humans, and on farms. (Viruses are not microbes because they do not have their own cell structure.) In the 1970s, when confinement systems for hogs emerged, farmers already had a method for dealing with infections—antimicrobials. Antimicrobials are chemical agents that either kill microbes or hinder their ability to grow and reproduce. (Antibiotics belong to a special class of antimicrobials that are effective at countering bacteria.) Commercially prepared antimicrobials have been available since 1935 and were simultaneously marketed for human health and for veterinary purposes beginning in 1938.[22] Very little has changed over the years regarding the human-animal speci- ficity of antimicrobials. In 2012, Page and Gautier identified twenty-seven different classes of antimicrobials that are commonly given to animals intravenously or as additives to food or water; of those, eighteen classes were also used in human medicine, and only nine were exclusively applied to animals.[23]

In 1949, very soon after introducing antimicrobials as a measure for livestock disease control, agrichemical researchers reported that animals treated with antimicrobials grew faster than those that were not treated. Weight gain or live-weight gain is the rate at which pigs grow. Measured in grams of body mass added per day per pig, it is a productivity measure for the farmer. More weight gained over a shorter time period increases profit so long as there is little or no increased cost to feed the pig. Feed efficiency, a cost measure, is the ratio of the weight gain of the pig to the weight of the food eaten by the pig over the same period of time. The agriculture industry noticed that feed efficiencies of antimicrobial-treated animals increased at the same time their weight gain increased.

Agricultural scientists have several working hypotheses regarding the mechanism for antimicrobial-induced growth promotion. These include that antimicrobials

- stave off low-level infections that consume the animal's resources and slow its growth;[24]
- reduce the number of intestinal microbes that consume nutrients from the animal's feed that would otherwise be used by the animal for its growth;
- reduce the number of intestinal microbes that produce metabolites that act to slow the growth of the animal;[25]
- thin the animal's intestinal walls so that they more quickly and efficiently absorb food nutrients than untreated animals; and
- promote bile acid production in the intestines, which prevents infection, spares nutrients for growth, and increases the animal's metabolic efficiency.[26]

Although it is still uncertain which, if any, of these hypotheses is correct, the results are significant even when animals are treated at subtherapeutic levels. Controlled studies of healthy pig populations in clean environments show 3.3–8.8 percent increases in weight gain and up to 7 percent increases in feed efficiency.[27] Presumably, better improvements are observed for animals living on working farms that have less-sterile environments. It is thus easy to understand why the use of antimicrobials in agriculture steadily grew from the 1940s through 2015. The US Centers for Disease Control conservatively reports, "It is difficult to directly compare the amount of drugs used in food animals with the amount used in humans, but there is evidence that more antibiotics are used in food production."[28]

The growth in antimicrobial use came with additional consequences. It was apparent as early as the 1960s that certain microbes were developing resistances to antimicrobials when the treatments were applied under the same circumstances in which they had previously been effective.[29] This so-called antimicrobial resistance develops as a normal evolutionary response to environmental pressures. Because a microbe is a living organism, it reproduces by manufacturing a replicate copy of its DNA. Sometimes the manufactured DNA isn't a perfect copy of the original, which results in an offspring microbe cell that—if it lives—is a mutated version of the original microbe cell. The mutated cell can have properties that help it resist the effectiveness of an antimicrobial agent. Mutations in DNA occur with predictable frequencies. Since generations of microbes come and go many times a day, they can quickly mutate and adapt to changes in their environment, including the pressure of antimicrobials.

Humans and all other animals pass on genes and genetic mutations through vertical gene transfer: the passing of genes directly from parents to offspring. But microbes have another powerful way of mutating that does not occur in normal human reproduction.[30] Called horizontal gene transfer or lateral gene transfer, there are three known mechanisms. In a process called conjugation, two bacteria can connect and transfer DNA strands between the two cells. Each cell may take on genetic properties that were formerly associated with the partner cell. Some stand-alone DNA strands, called plasmids, exist as separate entities from the cell's chromosomal DNA, but can still give the cell special useful properties, such as antimicrobial resistance. Because they are not part of the chromosomal DNA, they can easily be transferred from cell to cell. In a second process, called transduction, a bacterium can incorporate snippets of DNA from another bacterium into its own DNA. Viruses act as agents to gather the DNA snippets from one bacterium and deliver them to a second. The two virus-infected bacteria don't have to be of the same species, which means a virus can pass properties of antimicrobial resistance from one species of bacteria to another. In transformation, a bacterium can acquire pieces of DNA from the local environment (for example, the soil) and incorporate them into its own DNA.

Because microbes viably mutate at a rapid rate and because they have the power of horizontal gene transfer, it is careless to continuously administer the same antimicrobial with a regular implementation schedule in the same venue, as is commonly done in commercial animal agriculture. If one does, it's likely that eventually, whether through mutation or horizontal gene transfer, some hardy bacteria will survive the treatment, and those bacteria will leave behind living generations of hardy offspring in the CAFO and in the surrounding environment.

Since antimicrobials are not always fully processed by an animal's body, antimicrobials, and their metabolites, find their way into the surrounding environment. The primary vehicle of transfer is waste. A 100-kg hog produces roughly 6.3 kg of waste per day.[31] In a facility that houses (say) two thousand hogs, there's a lot of poop to deal with. The industry's solution is to collect the waste in open-air lagoons (holding ponds) until it can be spread on nearby farm fields. Guidry et al. describe the process of manure spreading:

> For disposal, the liquified hog waste is often sprayed as fertilizer on surrounding fields using conventional irrigation equipment. This aerosolizes the waste, creating a fine mist that can travel airborne for several miles depending on atmospheric conditions. The complex mixture of pollutants released during waste-spraying, lagoon off-gassing, and barn ventilation includes harmful gases such as ammonia, hydrogen sulfide, and methane; volatile organic chemicals, disinfectants, and cleaning agents; organic dusts containing dander, mold, and particulate matter; and viruses and bacteria, including antimicrobial-resistant types.[32]

Overspraying can result in runoff and the contamination of nearby waterways and groundwater.

Sometimes the lagoons leak or rupture, which, again, can result in water contamination. There were 2,057 reported manure spills in Iowa alone in the ten-year period from 2005 to 2015. At least 1,204 of these spills originated in hog farms. Unplanned and uncontrolled manure spills can spread anything that is in the manure—including heavy metals, antimicrobials, bacteria, and viruses.[33]

Antimicrobials can also be transferred into the wider environment via unused feed, by surreptitious birds and other wild animals, on the hands and clothing of the humans who tend the animals, through the air, and on the bodies and within the meat and internal organs of the animals that are sent to market.[34] Pathogenic bacteria and diseases can be transferred via these same routes, but there are fewer barriers in transferring antimicrobial resistance properties than there are in transferring pathogenic properties. First, pathogenic bacteria frequently cannot survive long outside a host, whereas any bacterium that lives in air, water, or on soil can have antibacterial-resistant properties. Second, pigs have complex immune systems designed for surmounting new doses of bacteria; environmental soil and water do not. Third, pigs and herds that harbor pathogenic bacteria are culled from the environment to terminate the spread of the disease, whereas antibacterial-resistant bacteria are left to thrive and spread resistance through both vertical and horizontal gene transfer.

The presence of high concentrations of antimicrobials within the proximity of a farm increases the likelihood that the local environment will have higher levels of antimicrobial resistance than the surrounding area. This can occur by the spreading of the unmetabolized antimicrobial itself, so that local bacteria develop antimicrobial resistance, or by the spreading of antimicrobial-resistant microbes or their genes into a new environment.[35] Indeed, several studies have not only reported higher levels of antimicrobial-resistant bacteria near farms than in the surrounding region, but they also find a greater-than-expected diversity of antimicrobial-resistant genes in the same bacteria samples.[36]

Higher levels of antimicrobial resistance can be shown to align well with the transmission routes of the antimicrobials. One study in South Africa showed that bacteria samples collected downstream from an urban center had higher levels of antimicrobial resistance than those samples collected upstream.[37] Another study showed that 87.5–98.5 percent of the bacteria strains isolated from groundwater at an urban site in Kenya had resistance to antibiotics used to treat intestinal sicknesses.[38] Ash et al. collected river-water samples downstream from city centers throughout the Midwestern United States and found that 44 percent of those bacteria that were resistant to two or more clinically important antibiotics were harboring DNA plasmids, and 70 percent of the detected plasmids carried a gene for resistance to ampicillin, a derivative of penicillin commonly administered to humans and farmed animals.[39] These results appear to indicate that waterways are effective mechanisms for spreading antimicrobial resistance away from centers of antimicrobial use.[40]

Antimicrobial resistance can be transferred to humans who regularly work with farmed animals.[41] The reverse transfer can also happen.[42] In 1976, Levy et al. reported that seven of eleven members of a farm family had acquired high levels of antimicrobial-resistant intestinal bacteria within six months of introducing antimicrobial supplement to the chickens' feed. (It took one week for the chickens to develop the same condition.) [43] Hatcher et al. report that children living in the same households as workers of industrial hog farms had higher rates of antimicrobial-resistant *Staphylococcus aureus* in their nasal mucous membranes than did other children.[44] They link these higher rates to the personal protective equipment that the workers bring home from hog farms. These are two of several studies that demonstrate how antimicrobial resistance can be fostered in the bacteria within farmed animals and later spread, by horizontal gene transfer, to the bacteria in humans who have come in direct contact with the animals, animal waste, or items associated with animal husbandry.[45]

The Clean Water Act (CWA) is the primary vehicle for federal oversight of CAFOs.[46] The Environmental Protection Agency (EPA) issued a CWA rule in 2003 that required all CAFOs to obtain a National Pollutant Discharge Elimination System permit unless they could demonstrate that there was no potential for discharge of pollutants into waterways. The permit limited the amount of permissible discharge and, in some cases, included a water-testing requirement. But a decision by the Second Circuit Court struck down the broad permit requirement, siding with industry groups who argued that the EPA could not require permits for merely potential polluters; permits could only be required for those who were *actually* discharging waste into waterways. In response, the EPA modified the rule so it included only those who were discharging or those who "propose to discharge," but this, too, was challenged and struck down in *National Pork Producers Council v. EPA*. The Fifth Circuit Court ruled that the "propose to discharge" standard is beyond the EPA's regulatory authority under the CWA.[47]

There is thus minimal federal oversight over CAFOs. States have the power to enact more stringent environmental requirements, but the regulatory climate in the Midwest, and particularly in Iowa, favors CAFOs. Iowa, in fact, passed a law in 2010 that prohibits state or county agencies from regulating CAFOs more stringently than does the federal government.[48]

Before turning to our argument, it is worth noting that the siting of CAFOs often raises issues of environmental justice. Hog CAFOs are often located in places where low-income communities and communities of color bear the brunt of the negative externalities associated with commercial animal farming.[49] Environmental injustice connected to CAFOs is most prevalent in the southern United States, but there are also documented cases in the Midwest. For example, in Ohio, CAFOs disproportionately impact black and Hispanic populations and low-income communities.[50]

Arguments for Regulation

In "Harm to Others: The Social Cost of Antibiotics in Agriculture," Jonny Anomaly argues in favor of the regulation of antimicrobial use in agriculture from the grounds of classical liberal political philosophy.[51] Anomaly argues that (at least one version of) the Harm Principle justifies limiting the liberty of producers by preventing them from using prophylactic antimicrobials (that is, antimicrobials used to prevent disease).[52] Anomaly's version of the Harm Principle permits state intervention in Agent A's liberty of action when "Agent A acts in a way that reduces B's welfare (especially by setting back B's interests), B does not consent to A's action, and harm to B is a predictable consequence of A's action."[53] Anomaly argues that nontherapeutic antimicrobial use in agriculture meets all of these conditions, since the emergence of antimicrobial-resistant bacteria is a predictable consequence of the practice, such bacteria have the potential to significantly set back individual interests, and those harmed did not consent to the use of antimicrobials in agriculture.

The Harm Principle originated in the work of John Stuart Mill. Mill's statement of the principle is usually interpreted as an invocation against paternalist laws and policies. It is a limit on state action: "That the only purpose for which power can be rightfully exercised over any member of a civilised community, against his will, is to prevent harm to others."[54] In theory, individuals may take whatever risks they like with their own well-being. Anomaly's version of the principle clarifies the notion of harm involved (setting back interests), introduces the condition of non-consent, and adds a predictability requirement. Nonetheless, the basic form of the principle remains the same. It is a limit on state action; as such, it provides necessary but not sufficient conditions for governmental intervention.[55] In classical liberal political philosophy, individuals are free to pursue their own interests in almost any way they see fit, consistent with the like freedom for others. The state may step in and limit their liberty *only when* the act or practice would result in harm to others. But since the Harm Principle provides only necessary conditions for governmental action, one needs an additional warrant for intervention. The Harm Principle tells us that if an act or practice does not cause harm to others, the state *may not* intervene, but it doesn't follow that the intervention is licensed in all cases of harm to others. Additional argument is needed to establish the normative claim that regulation ought to be enacted. Notably, if the harm of some regulation X outweighs the harm caused by the act or practice, then intervention is not justified.

The nature of the harm involved in the case of antimicrobial use in agriculture is also important. Each individual farmer's contribution to the overall harm of nontherapeutic antimicrobial use is minimal; it is the cumulative effect that causes the harm. Further, in most cases, the harm is *probabilistic*—it is merely the risk of harm, not actual harm. Finally, even in cases of demonstrable actual harm, the harm would be indirect. If Agent A steals from B, then there is a direct link between A's action and the reduction in B's welfare, and one can see how laws preventing theft pass the test of the Harm Principle. But there is no such link in the case of those who currently use nontherapeutic antimicrobials and those who have been or will (potentially) be harmed by the cumulative effect of antimicrobial use in agriculture. There is no specifiable person we can point to who commits the harm, nor is there a specifiable person we can point to who will be harmed.

Anomaly invokes Gerald Gaus's Public Harm Principle in response to these kinds of worries.[56] The Public Harm Principle is best interpreted as a way to assign responsibility in cases of cumulative harm. As such, it might address some of the worries noted above. But the Public Harm Principle remains a limit on governmental action and thus offers necessary but not sufficient conditions for state intervention. What Anomaly needs to demonstrate is that the harm of the use of antibiotics in agriculture outweighs the harm of regulation. Admittedly, this case could almost certainly be made. Even so, our contention is that there is a better way to approach the problem of nontherapeutic antimicrobial use in commercial animal agriculture. Before we turn to our argument, we need to introduce some terminology.

Economists characterize goods in a taxonomy that includes four distinct types. Goods are classified as one type or other depending on whether they are rivalrous and excludable.[57] A good is *rivalrous* in consumption if it is the case that if an agent A consumes a unit of the good, then A's consumption reduces the availability of the good. Consider, for example, the cupcakes at one's local bakery. The cupcakes are rivalrous because A's consumption of a cupcake reduces the number of cupcakes on the shelf. The ambiance at the bakery, on the other hand, is non-rivalrous. A's enjoyment of the ambiance does not reduce its availability.[58] A good is *excludable* if an agent can be prevented from consuming or accessing the good. A's cupcake can again serve as an example. Once A has purchased the cupcake, A can prevent others from consuming it. It is A's cupcake to consume. A's cupcake is a private good.

Private goods like A's cupcake are both rivalrous and excludable. Other private goods include things like cars, houses, apparel, stocks, bank accounts, and so on. Whenever A owns a particular private good, A's ownership right precludes others from owning or using it. Public goods are the opposite of private goods: they are non-rivalrous and non-excludable. Everyone can enjoy the good without depleting or diminishing it, and no one can be excluded from enjoying it. Examples of public goods include public roads, public parks, and free concerts. A's use of a city park does not deplete it, and everyone is welcome to use the park. Club (or toll) goods are non-rivalrous and excludable. As with public goods, a club good can be enjoyed without depleting or diminishing it, but, unlike public goods, individuals can be excluded from enjoyment of the good. Examples of club goods include toll roads and warehouse memberships. A's Costco membership does not diminish the availability of memberships, so the good is non-rivalrous. Since individuals can be excluded from enjoyment of membership, it is excludable. Finally, common-pool resource goods (sometimes just called common goods) are those that are rivalrous and non-excludable. An agent's use of the good depletes its availability, but no one can be excluded from access to the good. Examples of common-pool resource goods include aquifers and fisheries.

More abstract goods can be characterized in this taxonomy, too. For example, the herd immunity from pathogenic viruses provided by vaccinations is non-rivalrous and non-excludable and is thus a public good.[59] Everyone is protected by herd immunity, no one is prevented from enjoying it, and A's "consumption" of immunity does not preclude B from "consuming" it. Note that public goods often need public support—either through taxation or through participation. Herd immunity requires that enough individuals be vaccinated to protect everyone, including those who are not, and especially those who cannot be, vaccinated (for example, newborn infants and those whose immune systems are compromised).

We think that antimicrobial effectiveness is best characterized as a common-pool resource good. Prima facie, it may seem that antimicrobial effectiveness does not fit the definition of common-pool resource goods, since the drugs themselves are excludable. Given that pharmaceutical corporations own patents on particular antimicrobials, and individuals with bacterial infections must obtain a prescription for a drug and then purchase it, individuals can be excluded from access to antimicrobials. However, our concern is at the abstract level: we are concerned about the *effectiveness* of antimicrobials. Antimicrobial effectiveness is analogous in many respects to herd immunity. Vaccines themselves (at least those that are not in the public domain) are private goods, insofar as formulations and manufacturing processes are patented and owned by pharmaceutical corporations. The immunity they provide, though, is a public good. Everyone is protected (if enough people in the relevant community are immunized) and no one can be excluded. Likewise, we are all protected from bacterial infections by the effectiveness of antimicrobials (albeit after the fact). Thus, we suggest that antimicrobial effectiveness is non-excludable. Unlike herd immunity, though, antimicrobial effectiveness is rivalrous. If antimicrobial overuse engenders antimicrobial-resistant bacteria, then said overuse has depleted the resource so that it is unavailable to others. In this respect, the effectiveness of antimicrobials resembles fisheries and aquifers—if overfishing depletes the fishery, and overuse of water depletes the aquifer, then the resources those common goods provide are not available to others. Note that if antimicrobial protection were eternal, and evolution of antimicrobial-resistant bacteria were not possible, then antimicrobial effectiveness would be non-rivalrous. But that's not the case.

If we think of the protection that antimicrobial effectiveness affords the population as a whole as a good we share in common, then we can think of the overuse of antimicrobials as a problem of the commons. We thus suggest that we shift our focus away from attempting to justify the restriction of individual liberties and instead begin to think about the conditions under which it is permissible for individuals to use or appropriate goods held in common. The benefit of this shift is that it relocates the burden of proof. If our baseline is individual liberty (that we all may do as we wish as long as our actions do not cause harm to others) and we think of antimicrobials as private goods, then the burden of proof is on those who wish to argue for regulation. The presumption is that individuals have ownership rights over such goods, and state regulation must be justified against the backdrop of private property. If, instead, our focus is legitimate use or appropriation of goods held in common, then, we argue, the burden of proof shifts to those who wish to claim that use or appropriation is legitimate.

Many, if not most, common-pool resource goods are susceptible to overuse. Individual interests in attaining the good (or its products) often conflict with the collective interest in maintaining the resource. Of course, sometimes one may use or appropriate common-pool resource goods without ill effect, and in such cases, individuals should be free to use them. The question thus becomes: under what conditions is the regulation of the use of common-pool resource goods justified? We suggest that John Locke's theory of property can give us guidance here.[60] Locke argues that individuals may appropriate goods held in common only in cases where "enough and as good" is left for others.[61] This stipulation is often called the Lockean Proviso; it places limits on acquisition or use in cases where one's appropriation of resources held in common leaves others without access to the shared resource (or, at least, access to a resource "as good").[62]

In *The Limits of Lockean Rights in Property*, Gopal Sreenivasan argues that the Lockean Proviso is Locke's answer to "the consent problem."[63] Locke's goal in chapter 5 of *The Second Treatise of Government* is to legitimate individual appropriation of resources. When resources are held in common, individual appropriation appears to be unjust. If one could obtain the consent of the co-owners, of course, appropriation would be legitimate. But it is practically impossible to secure the consent of each and every co-owner when the resources in question are commonly owned by all. However, if "enough and as good" remains once one has appropriated that of which one can make use, then no one is harmed by the appropriation. As such, consent can be foregone.

However, in cases where a common-pool resource good is being depleted at such a rate that there will *not* be enough or as good left for others, then the burden of proof for use belongs to those who wish to appropriate or use the good. It is their burden, because consent to use or appropriate cannot be presumed. If Agent A overuses and depletes the stock of a particular common-pool resource good, then the co-owners of that good are left wanting. Appropriation in such cases is a form of unjust acquisition.[64]

Thus, we argue that the Lockean Proviso provides a guideline for when regulation of common-pool resource goods is warranted. Below, we argue that the nontherapeutic use of antimicrobials in agriculture violates the Proviso. Before turning to that argument, though, it is important to note the difference in this approach from that of Anomaly. Anomaly argued for regulation via the Harm Principle. The baseline for Anomaly's argument is individual liberty. As such, on the assumption that antimicrobials are private goods, the burden of proof is on those who wish to argue for regulation. The presumption is that individuals have ownership rights over them, and state regulation must be justified against the backdrop of private property. But if antimicrobial effectiveness is a common-pool resource good governed by the Lockean Proviso, and it is subject to overuse, then the burden of proof shifts to those who wish to claim that use or appropriation is legitimate.

We have seen that the effectiveness of antimicrobials is being depleted, and this depletion has been tied to uses in commercial animal agriculture. Antimicrobials—and the more abstract good of their effectiveness—are an essential tool in the arsenal of public health providers. As Martin et al. note, "The practice of medicine and the state of public health would be catastrophically affected if antibiotics were not generally effective in treating bacterial illnesses."[65] Enough and as good will not remain if we continue to use antimicrobials indiscriminately in agriculture, and so further use without consent is an unjust acquisition.[66] Regulation to prevent said acquisition is thus warranted. We suggest a ban on all nontherapeutic uses of antimicrobials in agriculture and strict regulation of therapeutic uses. It is in the collective interest to maintain the effectiveness of antimicrobials as long as possible. Nontherapeutic uses of antimicrobials in agriculture are nonessential, since there are other ways to prevent the spread of disease.

Consequences of Regulation

We cannot predict with certainty the consequences of the strict regulations we propose. However, the Iowa State University Extension Service provides guidelines for minimizing the use of antimicrobials in hog farms. The Service suggests that when the use of prophylactic antimicrobials is discontinued, biosecurity is of utmost importance. Farms must maintain "stringent controls on cleanliness and sanitation, animals entering the farm, feed quality, and environmental conditions to prevent or reduce stress (including transportation)."[67] The environmental recommendations designed to prevent or reduce stress include giving pigs more floor space, increasing the feeder space, and providing bedding.[68]

Denmark prohibited all nontherapeutic uses of antimicrobials in pigs effective 1999, so it serves as a good case study.[69] Note, though, that there are several differences between the kind of hog CAFO you're likely to find in the Midwest and one you will find in Denmark. Barry Estabrook, author of *Pig Tales: An Omnivore's Quest for Sustainable Meat*, describes the conditions on a hog CAFO he visited in Illinois: "[The] pigs are crowded in pens on hard slatted-floors that allow their excrement to fall into pits directly below their feet, where it stays for up to a year reeking and emanating poisonous gases that would kill the animals should the barns' ventilation fans fail."[70]

Sows, when pregnant, are kept in individual gestation crates that restrict movement. After a sow gives birth, she is transferred to a farrowing crate, which, again, leaves little room for movement. The piglets' tails are docked to prevent stress-related biting and their teeth are clipped or ground down to reduce the likelihood of injuries to other pigs.[71]

The conditions described above would not meet the farmed animal welfare standards in the European Union (EU). Pigs cannot be placed in individual stalls or crates in the EU. Further, farmers must provide pigs with material for rooting and playing; they cannot be housed in a space with merely a hard slatted floor. Tail docking is prohibited in the EU and tooth clipping and grinding are strongly discouraged.[72] The EU directive on the minimum standards for pigs requires that pigs "benefit from an environment corresponding to their needs for exercise and investigatory behavior" and notes that their welfare "appears to be compromised by severe restrictions of space." Finisher pigs must have at least 1 square meter of floor space per pig.[73] Thus, even before the antimicrobial ban, Danish hog farms were governed by EU animal welfare standards and were not as intensively crowded as those in the United States. Nonetheless, the prohibition on nontherapeutic uses of antimicrobials in Denmark resulted in several changes to the hog industry. Michael Nielsen, a Danish hog farmer, reported to the *New York Times* the kinds of changes he made after the law went into effect. To reduce stress, he increased the floor space available to each pig by 50 percent. He also built "safety" zones in stalls where sows give birth to provide piglets a place to sleep without the risk of being crushed by their mother.[74] Most Danish farmers have moved to a system in which piglets are kept with their mothers for roughly a month. This has two beneficial consequences: (1) It reduces the stress on piglets. Stress sometimes leads to infectious diarrhea. (2) It gives piglets time to build their immune systems naturally. When they are separated from their mothers too early, they are more susceptible to infection.[75]

All of these changes reduce the chances of infection and disease on hog farms. But they also impact the economic efficiency of hog operations. If you give pigs more floor space to reduce stress and decrease the likelihood of infection, you'll have to cut back on the number of pigs in your barn. If you allow piglets to nurse longer to build natural immunity, then you won't get as many litters per year per sow as you could otherwise. If you provide bedding, your farm becomes more labor intensive. As such, many U.S. hog farmers will likely be reluctant to implement the kinds of changes necessary to prevent or minimize disease without prophylactic antimicrobials. The *New York Times* reports that American pork industry officials interviewed after touring Danish pig farms said that implementing reforms similar to Denmark's would "markedly" increase pork prices.[76]

However, it is worth noting that Danish pig production has actually *increased* since the implementation of the ban on nontherapeutic antimicrobial use.[77] According to the Danish Agriculture & Food Council, roughly 28 million pigs are "produced" annually on 5,000 Danish pig farms.[78] Further, the Danish government is pushing to double the national annual pig production, from 28 million head to 56 million.[79]

One notable element that is essential to Denmark's success in eliminating nontherapeutic uses of antimicrobials in agriculture is the collaboration between the agricultural sector, veterinarians, human health researchers, and the Danish government.[80] Data on antimicrobial use was and remains readily available for each interested party to access. In fact, Denmark publishes its data annually in a publicly available document (DANMAP).[81] But things are very different in the United States. The only available data on U.S. antimicrobial use in agriculture is the Food and Drug Administration's nationwide sales figures for antimicrobials used in farmed animals. We don't know who is administering antimicrobials, the dosage amounts, or which animals are receiving them.[82] Further, U.S. agribusiness actively lobbies to prevent release of data on farm industry practices.

A controlled study conducted on an Irish hog farm confirms that prophylactic antimicrobials can be successfully eliminated. The researchers examined six groups of pigs from the weaner stage to slaughter. Each group contained 140 pigs. Half of the pigs were given in-feed antimicrobials, half were not. Individual pigs received therapeutic antimicrobials if and when they became sick or lame. At slaughter, those treated were roughly 2 kg heavier than those not treated, but the feed efficiency did not differ. The mortality rate at the finisher stage was higher for untreated pigs (3.13 percent vs. 2.14 percent).[83] Diana et al. note:

> Overall, withdrawal of prophylactic in-feed [antimicrobials] did not result in major detrimental problems for performance and health and welfare of pigs. Untreated pigs were as efficient as pigs fed with [antimicrobials] although there were numerical reductions in production performance and a tendency towards higher mortality in the finisher stage. These results indicate that the removal of

prophylactic in-feed [antimicrobials], while still allowing the use of parenteral [antimicrobials], is possible, but will need some extra measures to be implemented to avoid loss of profit and impaired pig welfare.[84]

The extra measures include adequate hygiene and better herd management. Indeed, the authors note that poor hygiene and high stocking densities facilitate the spread of disease and thus make the removal of prophylactic antimicrobials more challenging.[85] Note, too, that Irish hog operations are governed by the EU animal welfare standards, so the stocking density is lower than on typical hog farms in the Midwest.

Conclusion

We have argued that the effectiveness of antimicrobials is a common-pool resource good, subject to regulation. Strict regulations on the use of antimicrobials in animal agriculture should be implemented to prevent this essential resource from being depleted. The use of nontherapeutic antimicrobials in agriculture should be banned, and the use of antimicrobials for therapeutic purposes should be heavily regulated. The Danish and Irish examples demonstrate that this can be done successfully. Of course, there are several differences between EU farming practices and those in the United States. U.S. farms would almost certainly have to decrease the stocking density of farmed animals in CAFOs, which, in turn, will result in a smaller herd size.

There are, of course, negative consequences associated with the reduction in herd size. For example, it is likely that the price of pork products will increase. As we noted in the discussion of the Danish experience, American pork producers claim that implementing similar changes in the United States would markedly increase the price of meat.[86] Meat eaters would thus likely see their food expenses increase, and those on a limited budget could be harmed by such increases. Furthermore, pork producers themselves could be harmed by scaling back to a smaller herd size. In an industry where the profit margin is thin already, any reduction in profit could doom a producer.[87]

However, because antimicrobial resistance is a common-pool resource good, the burden is on producers to convince us that these costs justify the risks. Do these considerations meet that burden? We argue that they do not. We have seen that agricultural uses of antimicrobials have been tied to antimicrobial-resistant bacteria, and it is likely that cases like those we discussed are going to be more frequent in the future. While we ought to avoid harming those who are already worse off, there are ways to prevent such harm without continuing the practice of using antimicrobials prophylactically. Meat is not the only (or even the best) source of protein.[88] U.S. residents eat more meat than is healthy; the USDA estimates that the per capita average for meat and poultry consumption in 2018 was 101 kg (222 pounds).[89] A reduction in meat consumption is likely to have positive public health benefits.[90]

There are additional beneficial consequences of reducing herd sizes. Meat production itself carries with it a host of ills, most of which have a negative effect on human well-being (both individually and collectively). These ills are well-documented, so we will only briefly mention some of them here. As we noted earlier, waste is collected in manure lagoons, which sometimes rupture, and liquid manure can pollute the surrounding environment and contaminate groundwater. Nitrogen and phosphorus in the manure can lead to algae growth in contaminated waterways, choking off the oxygen and leading to dead zones where nothing can live. Excess nitrogen in drinking water can be hazardous to human health, especially for infants.[91]

Meat production also contributes an enormous amount of greenhouse gases to the environment, and raising livestock and processing meat requires vast amounts of water.[92] It is almost certainly the case that a reduction in the consumption of meat will have an overall positive impact on both individual and collective well-being by lessening the environmental impact of commercial animal agriculture.[93]

Finally, animal welfare benefits accompany the reduction in herd size as well. While hogs (and other livestock) will likely still be housed indoors and will probably still be unable to exhibit species-typical behaviors, a reduction in stocking density will at least give them more room.

Of course, getting regulations through a Congress heavily financed by agribusiness will be difficult, to say the least. Collecting data on antimicrobial use will also be extremely challenging. Implementing, enforcing, and overseeing the resultant regulations is yet another hurdle. Nonetheless, our recommendations are normative. Nontherapeutic uses of antimicrobials in agriculture ought to be banned, and the therapeutic use ought to be heavily regulated.

NOTES

1. E. B. White, *Charlotte's Web* (New York: Harper Collins, 1952), 242.

2. "Number of Operations by Size Group," Pork Checkoff, https://www.pork.org/facts/stats/structure-and-productivity/number-of-operations-by-size-group/.

3. USDA, "Quarterly Hogs and Pigs," National Agricultural Statistics Service (NASS), Agricultural Statistics Board, September 27, 2019.

4. Bjarne Peterson, "Dimension and Design of the Finisher Unit," *pig333.com* (blog), November 2, 2009, https://www.pig333.com/articles/dimension-and-design-of-the-finisher-unit_1977.

5. Brian Buhr, "Pig Space: Finding the Right Fit," *Pork: The Business Magazine for Professional Pork Production*, March 1, 2007.

6. David Jackson and Gary Marx, "Plan for 20,000 Hog Facility Sparks Revolt in Western Illinois," *Chicago Tribune*, December 28, 2016.

7. See the "Livestock Management Facilities Program," Illinois Department of Agriculture, https://www2.illinois.gov/sites/agr/Animals/LivestockManagement/Pages/default.aspx. Resolutions of county boards are nonbinding.

8. Jackson and Marx, "Plan."

9. Jeffrey Kittay, "In Illinois, a Victory against CAFO Construction," *The Counter*, January 27, 2017, https://thecounter.org/illinois-victory-cafo-construction/.

10. Janel M. Curry, "Care Theory and 'Caring' Systems of Agriculture," *Agriculture and Human Values* 19 (2002): 119–131.

11. Scott A. McEwen and Paula J. Fedorka-Cray, "Antimicrobial Use and Resistance in Animals," *Clinical Infectious Diseases* 34, no. 3S (2002): S93–S106.

12. McEwen and Fedorka-Cray, "Antimicrobial Use and Resistance."

13. "Timeline of FDA Action on Antimicrobial Resistance," FDA, current as of April 30, 2020, https://www.fda.gov/animal-veterinary/antimicrobial-resistance/timeline-fda-action-antimicrobial-resistance.

14. Scott Gottlieb, "Statement from FDA Commissioner Scott Gottlieb, M.D. on the FDA's 2017 Report on Declining Sales/Distribution of Antimicrobial Drugs for Food Animals, A Reflection of Improved Antimicrobial Stewardship," U.S. FDA, December 18, 2018, https://www.fda.gov/news-events/press-announcements/statement-fda-commissioner-scott-gottlieb-md-fdas-2017-report-declining-salesdistribution.

15. Michael J. Martin, Sapna E. Thottathil, and Thomas B. Newman, "Antibiotics Overuse in Animal Agriculture: A Call to Action for Health Care Providers," *American Journal of Public Health* 105, no. 12 (2015): 2409–2410.

16. Chris Dall, "FDA Reports Major Drop in Antibiotics for Food Animals," Center for Infectious Disease Research and Policy, December 19, 2018, http://www.cidrap.umn. edu/news-perspective/2018/12/fda-reports-major-drop-antibiotics-food-animals. Prophylactic antimicrobial use in cattle is still very high, too, for the same reasons.

17. Andrew Jacobs, "Denmark Raises Antibiotic-Free Pigs. Why Can't the US?" *New York Times*, December 6, 2019.

18. Thomas P. Van Boeckel et al., "Global Trends in Antimicrobial Use in Food Animals," *Proceedings of the National Academy of Sciences of the United States of America* 112, no. 18 (2015): 5649–5654.

19. Jonny Anomaly, "Harm to Others: The Social Cost of Antibiotics in Agriculture," *Journal of Agricultural and Environmental Ethics* 22 (2009): 423–435.

20. Leana Stormont, "Detailed Discussion of Iowa Hog Farming Practices," Michigan State University College of Law, Animal Legal & Historical Center, 2004, https://www. animallaw.info/article/detailed-discussion-iowa-hog-farming-practices. Farrowing is the process of producing piglets.

21. Caitlyn Elizabeth Abell, "Evaluation of Litters per Sow Year as a Means to Reduce Non-Productive Sow Days in Commercial Swine Breeding Herds and Its Association with Other Economically Important Traits" (master's thesis, Iowa State University, 2011), https://lib.dr.iastate.edu/cgi/viewcontent.cgi?article=3002&context=etd; Joe Vansickle, "Making 30 Pigs per Sow per Year a Dream Come True," *National Hog Farmer* (blog), January 15, 2009, https://www.nationalhogfarmer.com/genetics-reproduction/0109-producers-nearing-thershold.

22. C. Kirchhelle, "Pharming Animals: A Global History of Antibiotics in Food Production (1935–2017)," *Palgrave Communications* 4 (2018): 96.

23. S. Page and P. Gautier, "Use of Antimicrobials in Agriculture," *Revue scientifique et technique (International Office of Epizootics)* 31, no. 1 (2012): 145–188.

24. H. R. Gaskins, C. T. Collier, and D. B. Anderson, "Antibiotics as Growth Promotants: Mode of Action," *Animal Biotechnology* 13, no. 1 (2002): 29–42.

25. Metabolites are the intermediate or final products of biochemical reactions that serve to sustain a living organism. Alcohol is an example of a metabolite that is produced when yeast consumes sugar during the fermentation process. Metabolites can be the primary target products that are essential for healthy growth, side products that behave as secondary biochemical agents, or waste.

26. I. R. Ipharraguerre et al., "Antimicrobial Promotion of Pig Growth Is Associated with Tissue-Specific Remodeling of Bile Acid Signature and Signaling," *Science Reports* 8 (2018): 13671.

27. T. A. Van Lunen, "Growth Performance of Pigs Fed Diets with and without Tylosin Phosphate Supplementation and Reared in a Biosecure All-in All-out Housing System," *Canadian Veterinary Journal/La Revue Veterinaire Canadienne* 44, no. 7 (2003): 571–576.

28. U.S. Department of Health and Human Services, Centers for Disease Control, *Antibiotic Resistance Threats in the United States, 2013,* www.cdc.gov/drugresistance/threat-report-2013/pdf/ar-threats-2013-508.pdf.

29. Kirchhelle, "Pharming."

30. "Transfer of Antibiotic Resistance," Reactgroup.org, https://www.reactgroup.org/toolbox/understand/antibiotic-resistance/transfer-of-antibiotic-resistance/.

31. "Animal Manure Management," USDA Natural Resources Conservation Service, https://www.nrcs.usda.gov/wps/portal/nrcs/detail/null/?cid=nrcs143_014211#table1. For hogs and pigs, the total weight of manure produced is 63.1 pounds per 1,000 pounds of growing animal per day.

32. Virginia T. Guidry et al., "Connecting Environmental Justice and Community Health: Effects of Hog Production in North Carolina," *North Carolina Medical Journal* 79, no. 5 (2018): 324–328.

33. Margaret Carrel, Sean G. Young, and Eric Tate, "Pigs in Space: Determining the Environmental Justice Landscape of Swine Concentrated Animal Feeding Operations (CAFOs) in Iowa," *International Journal of Environmental Research and Public Health* 13, no. 9 (2016), https://doi.org/10.3390/ijerph13090849/. An additional 853 reports had unspecified sources (that is, the field was left blank), so it is reasonable to suppose that some of them were hog farms.

34. Christy Manyi-Loh et al., "Antibiotic Use in Agriculture and Its Consequential Resistance in Environmental Sources: Potential Public Health Implications," *Molecules* (Basel, Switzerland) 23, no. 4 (2018): 795. Several independent studies have reported measurable amounts of antibiotic residues contaminating chicken, beef, and milk products available in food markets in developing countries.

35. J. L. Martinez, "Environmental Pollution by Antibiotics and by Antibiotic Resistance Determinants," *Environmental Pollution* 157, no. 11 (2009): 2893–2902.

36. Manyi-Loh et al., "Antibiotic Use."

37. J. Lin, P. T. Biyela, and T. Puckree, "Antibiotic Resistance Profiles of Environmental Isolates from Mhlathuze River, KwaZulu-Natal (RSA)," *Water SA* 30 (2004): 23–28.

38. C. N. Wahome, "Contamination Levels of Groundwater, Antimicrobial Resistance Patterns, Plasmid Profiles and Chlorination Efficacy in Ongata Rongai, Kajiado North County, Kenya" (master's thesis, Kenyatta University, 2013).

39. R. J. Ash, B. Mauck, and M. Morgan, "Antibiotic Resistance of Gram-Negative Bacteria

in Rivers, United States," *Emerging Infectious Diseases* 8, no. 7 (2002): 713–716.

40. Manyi-Loh et al., "Antibiotic Use."

41. Bonnie M. Marshall and Stuart B. Levy, "Food Animals and Antimicrobials: Impacts on Human Health," *Clinical Microbiology Reviews* 24 (2011): 718–733.

42. J. F. Acar and G. Moulin, "Antimicrobial Resistance at Farm Level," *Revue Scientifique et Technique* 25 (2006): 775–792.

43. Stuart B. Levy, George B. FitzGerald, and Ann B. Macone, "Changes in Intestinal Flora of Farm Personnel after Introduction of a Tetracycline-Supplemented Feed on a Farm," *New England Journal of Medicine* 295 (1976): 583–588.

44. S. M. Hatcher et al., "The Prevalence of Antibiotic-Resistant *Staphylococcus aureus* Nasal Carriage among Industrial Hog Operation Workers, Community Residents, and Children Living in Their Households: North Carolina, USA," *Environmental Health Perspectives* 125, no. 4 (2017): 560–569.

45. Marshall and Levy, "Food Animals."

46. U.S. EPA, *Improving Air Quality: Eleven Years after Agreement, EPA Has Not Developed Reliable Emission Estimation Methods to Determine Whether Animal Feeding Operations Comply with Clean Air Act and Other Statutes Report*, No. 17-0396, September 19, 2017, https://www.epa.gov/sites/production/files/2017-09/documents/_epaoig_20170919-17-p-0396.pdf. Theoretically, CAFOs could be regulated under the Clean Air Act, too, but the Trump administration denied petitions seeking such regulations. The EPA's position is that it does not have enough information about CAFO emissions to issue regulations under the Clean Air Act.

47. Emily A. Kolbe, "Won't You Be My Neighbor? Living with Concentrated Animal Feeding Operations," *Iowa Law Review* 99 (2013): 415–448.

48. Kolbe, "Won't You Be My Neighbor?"

49. Kelly J. Donham et al., "Community Health and Socioeconomic Issues Surrounding Concentrated Animal Feeding Operations," *Environmental Health Perspectives* 115, no. 2 (2007): 317–320.

50. Julia Lenhardt and Yelena Ogneva-Himmelberger, "Environmental Injustice in the Spatial Distribution of Concentrated Animal Feeding Operations in Ohio," *Environmental Justice* 6, no. 4 (2013): 133–139.

51. Anomaly, "Harm to Others."

52. Recall that there are three kinds of uses of antimicrobials: *therapeutic* treatment of only diagnosed diseased animals, *metaphylactic* treatment of the herd after at least one animal has been diagnosed with a disease, and *prophylactic* treatment of the herd as a preventative to disease when there is no evidence of diseased animals. Anomaly also

discusses the use of antibiotics of growth-promoters, but since the FDA has banned that practice, we largely ignore their use in that capacity in this paper.

53. Anomaly, "Harm to Others," 425–426.

54. John Stuart Mill, *On Liberty* (Kitchener, ON: Batoche Books, 2001), 13.

55. Anomaly is aware of this and offers an additional argument to close the gap. This argument is discussed below.

56. See Anomaly, "Harm to Others," 430–432. See Gerald F. Gaus, *Social Philosophy* (London: M.E. Sharpe, 1999), chapter 8, for discussion of the Public Harm Principle.

57. The following discussion is based on Elinor Ostrom, Roy Gardiner, and James Walker's *Rules, Games, and Common-Pool Resources* (Ann Arbor: University of Michigan Press, 1994).

58. Of course, there is a point at which the bakery would be at capacity and thus the ambiance temporarily unavailable to others. But when A (or another patron) leaves, the ambiance remains for others to enjoy.

59. "Herd Immunity and COVID-19: What You Need to Know," Mayo Clinic, June 6, 2020, https://www.mayoclinic.org/diseases-conditions/coronavirus/in-depth/herd-immunity-and-covid-19/art-20486808. Herd immunity requires that a significant portion of the population be immune to the disease. Immunity is achieved either through vaccination or natural infection and recovery. Once a certain threshold of immunity is reached, the disease can no longer spread through the population. Note that the threshold is higher in diseases that are highly contagious. For example, the threshold for measles is 94 percent.

60. See John Locke, *Two Treatises of Government*, with intro. and notes by Peter Laslett (New York: Cambridge University Press, 1963), 2nd Treatise, chapter 5.

61. There are three conditions that must be met for legitimate property acquisition on a Lockean theory of property: (1) the spoilage limitation; (2) the "enough and as good" proviso; and (3) the charity limitation. The spoilage limitation is no longer in play once money enters the picture, so we do not discuss it here. The charity limitation is not directly relevant to our discussion. See Locke, *Two Treatises*.

62. There is disagreement among Locke scholars over the force of the Proviso after the advent of civil society. Those who think the Proviso remains in effect in civil society and can limit the acquisition of common goods include Robert Nozick, *Anarchy, State, and Utopia* (New York: Basic Books, 1974); James Tully, *A Discourse on Property: John Locke and His Adversaries* (Cambridge: Cambridge University Press, 1980); A. John Simmons, *The Lockean Theory of Rights* (Princeton, NJ: Princeton University Press, 1992); and Gopal Sreenivasan, *The Limit of Lockean Rights in Property* (New York: Oxford,

1995). Those who think it does not remain in effect include C. B. Macpherson, *The Political Theory of Possessive Individualism: Hobbes to Locke* (Oxford: Clarendon Press, 1962); and Jeremy Waldron, *The Right to Private Property* (Oxford: Clarendon Press, 1988). Waldron argues that the Proviso is a sufficient (but not necessary) condition for appropriation, even in the state of nature.

63. Sreenivasan, *The Limit.*

64. See Sreenivasan, *The Limit*, chapter 6, for a discussion of the appropriation of land when so many are without the means to support themselves as a form of theft.

65. Michael J. Martin, Sapna E. Thottahil, and Thomas B. Newman, "Antibiotic Overuse in Animal Agriculture: A Call to Action for Health Care Providers," *American Journal of Public Health* 105, no. 12 (2015): 2409.

66. One might question whether the Proviso, as a criterion of sustainability, is sufficient to support our argument. But note that there is no substitute for antimicrobial effectiveness; there is nothing even reasonably close to "as good." Thus, even on a weak sustainability reading of the Proviso (where the "enough and as good" condition is satisfied as long as there is a suitable substitute for the resource in question), maintaining antimicrobial effectiveness for as long as possible is the only way to meet the Proviso. Maintaining effectiveness can thus be seen as functionally equivalent to "enough and as good" in this particular case. We thank an anonymous reviewer for raising this question.

67. Mark Honeyman et al., "Minimizing the Use of Antibiotics in Pork Production," Iowa State University Extension, October 2002, https://store.extension.iastate.edu/product/2404/, p. 3.

68. Honeyman et al., "Minimizing," 8. They note that "the effectiveness of bedding in modifying the environment is dependent on the bedding quality that can be compromised if improperly harvested or stored."

69. Sharon Levy, "Reduced Antibiotic Use in Livestock: How Denmark Tackled Resistance," *Environmental Health Perspectives* 122, no. 6 (2014): A160–A165.

70. Barry Estabrook, *Pig Tales: An Omnivore's Quest for Sustainable Meat* (New York: W.W. Norton & Co., 2015), 20.

71. Estabrook, *Pig Tales*, 30. See "Farm Animal Confinement Bans by State," ASPCA, https://www.aspca.org/animal-protection/public-policy/farm-animal-confinement-bans. Gestation crates are prohibited or being phased out in twelve states: Arizona, California, Colorado, Florida, Kentucky, Maine, Massachusetts, Michigan, Ohio, Oregon, Rhode Island, and Washington. They are still permitted in most of the Midwest.

72. "Pigs," European Commission, https://ec.europa.eu/food/animals/welfare/practice/farm/pigs_en.

73. Council of the European Union, "Council Directive 2008/120/EC of 18 December 2008 Laying Down Minimum Standards for the Protection of Pigs," https://eur-lex.europa.eu/legal-content/EN/TXT/?uri=CELEX:32008L0120/. See the table for specific space requirements.

74. Andrew Jacobs, "Denmark Raises Antibiotic-Free Pigs. Why Can't the US?" *New York Times*, December 6, 2019.

75. Levy, "Reduced Antibiotic Use."

76. Jacobs, "Denmark."

77. Levy, "Reduced Antibiotic Use."

78. "Danish Pig Meat Industry," Danish Agriculture & Food Council, 2019, https://agricultureandfood.dk/danish-agriculture-and-food/danish-pig-meat-industry.

79. Kjeld Hansen, "Danish Bacon: What Happens When You Push Pigs to the Limit?" *The Guardian*, November 30, 2019.

80. Levy, "Reduced Antibiotic Use."

81. DANMAP, https://www.danmap.org.

82. Levy, "Reduced Antibiotic Use."

83. Alessia Diana et al., "Removing Prophylactic Antibiotics from Pig Feed: How Does It Affect Their Performance and Health?" *BMC Veterinary Research* 15, art. 67 (2019), https://doi.org/doi:10.1186/s12917-019-1808-x.

84. Diana et al., "Removing Prophylactic Antibiotics."

85. Diana et al., "Removing Prophylactic Antibiotics."

86. Bernard Rollin, "Antibiotic Use and the Demise of Husbandry," *Journal of Ethics* 22 (2018): 45–57. Rollin contests this claim by the meat industry. He cites a figure of $10 per year per family rise in cost if we banned subtherapeutic uses of antibiotics. Rollin quotes a 1999 study by the National Research Council Committee on Drug Use in Food Animals (available at https://www.ncbi.nlm.nih.gov/pubmed/25121246), but the figure is almost certainly higher now, twenty years later.

87. For a discussion of margins, see Chip Whalen, "Impact of Feed Costs on Hog Margins," National Hog Farmer, February 28, 2022, https://www.nationalhogfarmer.com/marketing/impact-feed-costs-hog-margins. The costs involved in raising pigs fluctuates, depending on the costs of inputs (for example, feed, veterinarian visits); the price the producer receives for each pig also fluctuates.

88. See "Protein," Harvard T.H. Chan School of Public Health, https://www.hsph.harvard.edu/nutritionsource/what-should-you-eat/protein/.

89. Keithly Jones, Mildred Haley, and Alex Melton, "Per Capita Meat and Poultry Disappearance: Insights into Its Steady Growth," USDA: Economic Research Service, June 4, 2018, https://www.ers.usda.gov/amber-waves/2018/june/per-capita-red-meat-and-poultry-disappearance-insights-into-its-steady-growth/. The United States has the highest annual average consumption of meat and poultry of any country. Only the U.S., Australia, New Zealand, and Argentina exceed 100 kg per capita. See Max Roser and Hannah Ritchie, "Our World in Data: Food Supply," United Nations Food and Agricultural Association, https://ourworldindata.org/food-supply#all-charts-preview.

90. Andrew Joyce et al., "Reducing the Environmental Impact of Dietary Choice: Perspectives from a Behavioural and Social Change Approach," *Journal of Environmental and Public Health* (2012), https://doi.org/10.1155/2012/978672. For example, a reduction in meat consumption is a recommended strategy to "reduce the high rates of some chronic diseases such as cardiovascular disease and certain cancers."

91. "Nutrient Pollution," EPA, https://www.epa.gov/nutrientpollution/issue.

92. See Rollin, "Antibiotic Use," for a fairly comprehensive discussion of the negative effects of conventional meat production.

93. For discussion, see the Intergovernmental Panel on Climate Change, *Climate Change and Land: An IPCC Special Report on Climate Change, Desertification, Land Degradation, Sustainable Land Management, Food Security, and Greenhouse Gas Fluxes in Terrestrial Ecosystems*, 2019, https://www.ipcc.ch/srccl/.

SUGGESTED READINGS

Duckenfeld, Joan. "Antibiotic Resistance Due to Modern Agricultural Practices: An Ethical Perspective." *Journal of Environmental Ethics* 26 (2013): 333–350.

Harfeld, Jes Lynning. "Telos and the Ethics of Animal Farming." *Journal of Agricultural and Environmental Ethics* 26, no. 3 (2013): 691–709.

Hoffman, Steven J., et al. "How Law Can Help Solve the Collective Action Problem of Antimicrobial Resistance." *Bioethics* 33, no. 7 (2019): 798–804.

Holthaus, Gary. *From the Farm to the Table: What All Americans Need to Know about Agriculture.* Lexington: University Press of Kentucky, 2009.

Ilea, Ramona Cristina. "Intensive Livestock Farming: Global Trends, Increased Environmental Concerns, and Ethical Solutions." *Journal of Agricultural and Environmental Ethics* 22, no. 2 (2009): 153–167.

Imhoff, Daniel, ed. *The CAFO Reader: The Tragedy of Industrial Animal Factories.* Berkeley, CA:

Watershed Media, 2010.

Kimbrell, Andrew, ed. *The Fatal Harvest Reader: The Tragedy of Industrial Agriculture*. Washington, DC: Island Press, 2002.

Mathew, Alan G., Robin Cissell, and S. Liamthong. "Antibiotic Resistance in Bacteria Associated with Food Animals: A United States Perspective of Livestock Production." *Foodborne Pathogens and Disease* 4, no. 2 (2007). https://doi.org/10.1089/fpd.2006.0066.

McWilliams, James. "The Ethics of Humane Animal Agriculture." In *The Routledge Handbook of Food Ethics*, edited by Mary C. Rawlinson and Caleb Ward. Abingdon, UK: Routledge, 2016.

Rollin, Bernard. "Ethics, Science, and Antimicrobial Resistance." *Journal of Agricultural and Environmental Ethics* 14, no. 1 (2001): 29.

Rossi, John. "Industrial Farm Animal Production: A Comprehensive Moral Critique." *Journal of Agricultural and Environmental Ethics* 27, no. 3 (2014): 479–522.

Sperling, D. "Food Law, Ethics, and Food Safety Regulation: Roles, Justifications, and Expected Limits." *Journal of Agricultural and Environmental Ethics* 23 (2010): 267–278.

Thompson, Paul B. *From Field to Fork*. New York: Oxford University Press, 2015.

Twine, Richard. "Animals on Drugs: Understanding the Role of Pharmaceutical Companies in the Animal-Industrial Complex." *Journal of Bioethical Inquiry* 10, no. 4 (2013): 505–514.

Ventola, C. Lee. "The Antibiotic Resistance Crisis: Part 1: Causes and Threats." *P & T: A Peer-Reviewed Journal for Formulary Management* 40, no. 4 (2015): 277–283.

Williams-Jones, Bryn. "Managing Antimicrobial Resistance in Food Production: Conflicts of Interest and Politics in the Development of Public Health Policy." *Les Ateliers de l'éthique* 5, no. 1 (2010): 156–169.

Prairie Dog Wars, the Philosophy of Biology, and Justice Scalia

Ian A. Smith

A champion of restoring ecological landscapes in Kansas to a time before European settlers came to America, Larry Haverfield decided to bring back the prairie dog to his property in western Kansas. Prairie dogs were abundant in the Midwest at the turn of the twentieth century; in fact, it is estimated that black-tailed prairie dogs (*Cynomys ludovicianus*), the most numerous of the prairie dog species, occupied at least 40 million hectares at the time.[1] Around the same time, ranchers argued that prairie dogs were causing serious problems for them, including consuming forage that is otherwise eaten by cattle and horses.[2] Such ranchers were successful in getting legislation passed in various states that aimed to eradicate the prairie dog from the Midwest. The laws have been largely successful, as prairie dogs have been eliminated from over 95 percent of their historic range.[3] One such law is Kansas Statutes Annotated (KSA) 80–1202, passed in 1909, which allows township trustees in towns of Kansas to direct the eradication of all prairie dogs within their respective townships. It also requires the cooperation of landowners in eradicating the prairie dogs; the trustees or their agents have the right to enter the lands of uncooperative owners and poison prairie dogs, even *without* the owners' consent. And then the owners are required to pay for the eradication, backed up by a lien upon their real estate if the owners fail to do so![4]

The township trustees of Logan County decided to act on KSA 80–1202 by threatening to enter the Haverfield Complex, a ranch owned by Gordon Barnhardt and the Haverfields (Larry and his wife Elizabeth), to exterminate the prairie dog "infestation," as the law refers to the habitation of prairie dogs in Kansas. The only problem with this plan was the presence of a not insignificant creature, the black-footed ferret. In wishing to restore the ecological landscapes of Kansas and in wishing to control the prairie dog populations as had been done prior to European settlement, Haverfield and the United States Fish and Wildlife Service (FWS) cooperated in 2007 to reintroduce the endangered black-footed ferret.[5] The black-footed ferret is the only federally listed endangered species that has been reintroduced to Kansas. And this species had not seen Kansan soil since the mid-twentieth century. In a word, this reintroduction was significant.[6]

To speak to the importance of the relationship between the ferret and the prairie dog, the black-footed ferret cannot survive without the prairie dog. About thirty-five thousand years ago, the black-footed ferret developed such a specialized relationship with the prairie dog that the black-footed ferret became fully separate from its generalist polecat predecessor.[7] What constitutes this highly specialized relationship is that over 90 percent of the diet of the black-footed ferret is the prairie dog, and the black-footed ferret uses the underground prairie dog towns as shelter.[8]

The District Court for Logan County prevented the township trustees of Logan County from carrying out the planned eradication.[9] As a federally listed species, the black-footed ferret is a protected species under the Endangered Species Act (ESA).[10] And when a state statute like KSA 80–1202 is in conflict with a federal statute, the federal statute trumps the state statute, per the Supremacy Clause of the U.S. Constitution, which requires that any federal law that is in conflict with a state law renders the state law invalid (or not legally binding).[11] The plan of the trustees to kill all prairie dogs in the Haverfield Complex would clearly result in the death of the approximately fifty ferrets that also lived on the land, as the ferrets would then have lost their prey. On appeal, representatives of the trustees challenged the ruling of the district court to the Court of Appeals of Kansas.

In 2012, the Court of Appeals affirmed the district court's decision and so ended the extermination plans of the trustees.[12] In their reasoning, the court referred to two pieces of the ESA that are particularly relevant to this chapter. According to the ESA, it is unlawful for any person subject to the jurisdiction of the United States to . . . take any such [endangered] species within the United States or the territorial sea of the United States. **16 U.S.C. §1538(a)(B)** [U.S.C. stands for United States Code, the compilation of the federal statutes of the United States].[13]

And, on the ESA:

> The term "take" means to harass, harm, pursue, hunt, shoot, wound, kill, trap, capture, or collect, or to attempt to engage in any such conduct. **16 U.S.C. §1532(19)**.

The FWS further defines "harm" and "harass" within the definition of "take" (but does not define the other terms, a point that will become relevant later). The FWS defines "harm" as any act that results in

> "significant habitat modification or degradation where it actually kills or injures wildlife by significantly impairing essential behavioral patterns, including breeding, feeding, or sheltering. **50 CFR 17.3(c)** [CFR stands for Code of Federal Regulations, the compilation of regulations promulgated by federal agencies and departments of the United States].

The regulation 50 CFR 17.3(c) is statutorily linked to §1532(19) in the following sense: when the ESA mentions "harm" in its definition of "take," the way the FWS defines it is the way it is to be understood in the ESA. If the trustees decided to exterminate all prairie dogs within the complex, that would significantly impair the essential feeding and sheltering patterns of the protected black-footed ferrets. Hence, the proposed extermination of the prairie dog would harm and so take the endangered black-footed ferret under 50 CFR 17.3(c) and §1532(19), thus violating section §1538(a)(B) of the ESA.

In this chapter, I wish to challenge the consistency of the FWS definition of "harm" and the ESA. In this challenge, we'll see how the philosophy of biology more generally and the philosophy of taxonomy more specifically can help in the interpretation of statutes like the ESA. The field of statutory construction or interpretation asks questions such as the following: Does a statute always have a "plain meaning" that is clear to those to whom it is directed? Is the meaning of a law something that is set when the statute is passed by Congress, or can the meaning of a law change over time? Other questions include: Is a law's meaning determined by the intentions of the legislators who voted for it, or by the purposes the legislature claims the law was meant to address, or by what the reader or interpreter of a statute understands it to mean?[14]

In this chapter, I lay out other relevant parts of the ESA, along with explaining what the reader needs to know about species and about certain philosophical issues about species. I investigate how the late stalwart conservative Justice Antonin Scalia interpreted the fit between the FWS's definition and the ESA in a landmark Supreme Court case, *Babbitt v. Sweet Home*.[15] I also consider objections to Scalia's argument that the FWS definition of "harm" is inconsistent with the ESA. Here, I come to Scalia's defense (oh my, how could a liberal environmental philosopher such as myself do such a thing!) by responding to said objections. In my response, I will further fill out his argument and appeal to what I have written elsewhere about whether we harm individual nonhuman organisms or nonhuman species when we impair breeding behaviors in our various human-habitat expansion activities.[16] This question is related to the following broader question prominent in the field of environmental ethics: whether it makes sense to say that we can harm populations or even species understood as meta-populations.

It may sound like I wish to challenge the ESA itself. Nothing could be further from the truth, for the motivation for challenging the consistency of the FWS definition of "harm" and the ESA is precisely to clean up what I view as a faulty link between the definition and the ESA. It is this faulty link that could get in the way of interpreting the ESA in such a way that it carries out the purpose of the ESA, as laid out in §1531(b): "the purposes of this chapter [of the United States Code] are to provide a means whereby the ecosystems upon which endangered species and threatened species depend may be conserved, to provide a program for the conservation of such endangered species and threatened species."

Species and the ESA

The question of what a species is happens to be a difficult inquiry (to say the least), especially since there are two ways of asking this question: metaphysical and biological. The so-called species problem has been an intractable problem in the philosophy of biology literature for a long time now, which is partly due to the conflation of different ways of asking this question.

The metaphysical question will not concern us here, but the biological one will.[17] The biological question can be further divided into two distinct subquestions. The one that I will consider here is this: By virtue of what are species distinct from higher-order biological taxa, that is, taxa above the species level? (The plural "taxa" are groups in a biological classification [or system], whereas the singular, a "taxon," is one grouping within that classification [or system].)[18] On the Linnaean hierarchy—the hierarchy we all learned about in high school—this is to inquire about the nature of the species category (or rank) as a taxonomic category, and how it differs from the genus category (or rank), for example. On the modern version of the Linnaean hierarchy, the major ranks in the hierarchy are these, starting from the most inclusive rank: domain, kingdom, phylum, class, order, family, genus, and species. Or, on the competing hierarchy of cladism, this is to inquire about the nature of the species category and how it differs from the other kind of entity that constitutes the tree of life, clades. A clade can be thought of as a group of branches in the tree of life, and a species can be thought of as one of the branches. In other words, in answering this question about species we are defining species in general, irrespective of some particular species that might be referred to by someone. That is, here we are understanding what the term "species" *means*; here we are developing a species concept.[19]

Now that we have a sense of what the biological question is that is useful for us, we can further understand how taxonomy can help with our present discussion. The first biological entity that is typically understood to exist below the species level is the subspecies. I won't define "subspecies" here, for whether subspecies are even real is controversial, and because we won't be concerned with subspecies as such. The next entity below that of the subspecies is the population. In fact, species are typically understood to be meta-populations, as in groupings of populations, where the grouping criterion in question will depend on the specific species concept used. A population can be defined in many ways, but let's consider how 50 CFR 17.3 defines it, as how that defines it is statutorily linked to the ESA: "A population is a group of fish or wildlife in the same taxon below the subspecific level, in common spatial arrangement that interbreed in nature." Note that this definition is consistent with what I have said above about the typical way these entities are understood.

Populations are made up of individual biological organisms, in which an individual organism is understood to be a *member* of or a *part* of a species.[20] It is this which another part of the ESA relevantly refers to in §1532(8), as the FWS definition of "harm" refers to wildlife: "the term 'fish or wildlife' means any member [or part] of the animal kingdom." So to review, organisms make up populations; on the next taxonomic level the subspecies is found (supposing we understand subspecies to be real); the next, the species; and then on the next we have the genus (on the Linnaean hierarchy) or the clade (on cladism).

Of course, things may not be this hierarchically clean. For example, there may be one last remaining population of the species. In this case, the species just reduces to the (last remaining) population. Perhaps this is at least partly what Congress had in mind when they defined "species" to include the possibility of just one population of species being a species, for Congress in the ESA defines the term §1532(16) thusly: "The term 'species' includes any subspecies of fish or wildlife or plants, and any distinct population segment of any species of vertebrate fish or wildlife which interbreeds when mature." This definition is surprisingly inclusive and encompasses even those biological entities that are typically understood to exist below the species level.

Scalia's Criticism of the Legality of the FWS Definition of Harm

The central question in *Babbitt v. Sweet Home* was whether the secretary of the Interior at the time, Bruce Babbitt, exceeded his authority under the ESA in promulgating the FWS definition of "harm" in the definition of "take." Or, in words that a philosopher like me is familiar with, whether the FWS definition is consistent with the ESA. Philosophers are almost obsessively concerned with consistency. But perhaps judges interpreting laws could be said to be equally concerned. Sir Edward Coke in 1628 remarked about the canon of statutory construction, the Whole Text Canon, which says that the text must be interpreted as a whole, taking the logical relation of its many parts into consideration: "It is the most natural and genuine exposition of a statute to construe one part of the statute by another part of the same statute, for that best expresseth the meaning of the makers."[21] Granted, in this case, we are seeing whether a piece of the Code of Federal Regulations is consistent with the ESA, a federal statute, but the importance of consistency is still the same, as how we understand "harm" in the ESA is supposed to be spelled out by what the CFR says harm is. What the Supreme Court ruled in this case is that Babbitt did not exceed his authority, but Antonin Scalia offered an incredibly logically nuanced dissent, one that assumed the importance of the Whole Text Canon. I won't go over his whole dissent here, of course (that would tire even the most diligent reader); rather, I will review the most salient aspects of his dissent for our purposes in this chapter.

Central to Scalia's dissent was another doctrine of statutory construction that can be derived from the Whole Text Canon, namely, the Associated-Words Canon (or *noscitur a sociis*), which generally means that a "word is known by the company it keeps."[22] More specifically, when "several items in a list share an attribute [that fact] counsels in favor of interpreting the other items as possessing that attribute as well."[23] To provide some examples, *noscitur a sociis* was referred to in the 2020 Impeachment Trial of President Donald J. Trump. Alan Dershowitz was counsel to Trump during the trial, and during his defense of Trump, Dershowitz relayed an example that Scalia himself had once used in the past: "If one speaks of Mickey Mantle, Rocky Marciano, Michael Jordan, and other great competitors, the last noun does not reasonably refer to Sam Walton, who was a great competitor but in business, or to Napoleon, a great competitor on the battlefield."[24] Rather, it would only reasonably refer to great sports competitors. Dershowitz then rehearsed an argument from then-retired Justice Benjamin Curtis, who acted as counsel for President Andrew Johnson when Johnson was undergoing his own impeachment trial. This argument was specifically in defense of Johnson against certain articles of impeachment that did not rise to the level of impeachable offenses, in Curtis's view. Judge Curtis applied *noscitur a sociis* to the following group of words: "treason, bribery, and other high crimes and misdemeanors," the language found in the United States Constitution laying out rules of the impeachment of presidents.[25] Here is the larger context from which those words are taken: "The President, Vice President and all civil Officers of the United States, shall be removed from Office on Impeachment for, and Conviction of, Treason, Bribery, or other high Crimes and Misdemeanors." Curtis claimed that "other high crimes and misdemeanors" should be interpreted to indicate only serious criminal behavior, akin to treason and bribery. The quality or attribute of commonality here is thus serious criminal behavior.[26]

To review our case at hand, recall the definition of "take" in the ESA: to take "means to harass, harm, pursue, hunt, shoot, wound, kill, trap, capture, or collect, or to attempt to engage in any such conduct (§1532(19))." Scalia identifies that the quality of commonality in "pursue, hunt, shoot, wound, kill, trap, capture, or collect" is that these terms *only* apply to affirmative conduct intentionally directed at individual organisms, rather than to anything higher in our taxonomic classification or system, such as a population.[27] I won't bother here with his argument about the "affirmative conduct intentionally directed" piece, as that won't be my focus. My focus is his point that these terms can only apply to individual organisms rather than to anything higher in our taxonomic system, given our question of how our knowledge of biology and the philosophy of biology can help in statutory construction. The question is then whether the quality of commonality in "harass" and "harm" is also supposed to be that these terms would only apply to individual organisms. We need not consider "harass" here, as that is not in question (though the reader should know that "harass" is also defined in 50 CFR 17.3). So, what is in question is whether the quality of commonality in "harm" is that this term only applies to individuals. Recall that *noscitur a sociis* means that if several items in a list share a feature, that sharing should lead us to interpret the other items as possessing that feature too. Since the following terms: "pursue," "hunt," "shoot," "wound," "kill," "trap," "capture," or "collect" all only apply to individual organisms, then so does "harm."

However—and this is the crucial point—Scalia argues that the FWS definition of "harm" in 50 CFR 17.3(c) not only refers to individuals, but it also refers to populations. Let's reconsider the definition of "harm": any act that results in "significant habitat modification or degradation where it actually kills or injures wildlife by significantly impairing essential behavioral patterns, including breeding, feeding, or sheltering." Specifically singling out breeding behavior, Scalia argues that impairing breeding behavior can *only* harm a population (or higher-level taxonomic unit) rather than an individual. This is because Scalia claims that one cannot injure an individual animal by impairing its breeding behavior; at most, one can only injure a population of animals by impairing breeding behavior. Perhaps one could be said to injure a population because by impairing breeding behavior, fewer individuals are coming into existence than if the impairment had not occurred. And if breeding behavior can only harm a population in Babbitt's definition (though he didn't see that, according to Scalia), and if "harm" can only apply to individual organisms in the ESA, then Babbitt's definition of "harm" is inconsistent with the ESA, as the ESA has it that "harm" in the definition of "take" can only apply to individual organisms.

What I intend to do now is fill out Scalia's argument for his identification of the quality of commonality in "pursue," "hunt," "shoot," "wound," "kill," "trap," "capture," or "collect" being that these terms *only* apply to individual organisms, rather than to populations. This is because it wasn't obvious to me that this was the case, and I imagine that it won't be obvious to other readers that this is the case. In this expansion of his argument, I will refer to how the OED (*Oxford English Dictionary*) defines these terms, as the OED is considered the definitive dictionary of the English language.[28] This expansion of his argument will turn out to be consistent with his statutory constructive method of semantic textualism. Broadly, a textualist interprets a statute based upon what the legislators in question intended in enacting a statute. This is opposed to a strict constructionist view where statutes are interpreted based on precisely what the words in the statute literally mean, and nothing else. In other words, a strict constructionist would not look at the context in which the words were used; rather, a strict constructionist could merely consult a dictionary to determine what words mean in a statute.[29] An adherent of semantic textualism holds that a statute is supposed to be interpreted based upon what the legislators intended to *say* in using the language they chose when enacting the statute, so a semantic textualist would look at what words mean but also their context.[30] This is opposed to an expectation textualist, who holds that a statute is supposed to be interpreted based on what the legislators intended or expected would be the *consequence* of their using the language they chose.[31]

I'll start with "hunt," as it uses "pursue" and as it shows up in the OED's definition of "pursue." This is not ideal, of course, as both terms are being defined in terms of each other, but I don't think that this violation of definition construction is so bad that it will prevent us from understanding the two terms. I'll then move on to "pursue" and move down the list. Starting with "hunt" then, one does not hunt a population, for the target of a hunt is not a whole population; rather, it is (an) individual(s). Why? Let's consider the relevant transitive definition of the verb "to hunt": to pursue (wild animals or game) for the purpose of catching or killing; to chase for food or sport; (often specifically) to pursue with hounds or other tracking beasts. By relevant, I mean the meanings of such terms as they would be ordinarily used in context with biological entities.[32] And now for the intransitive verb "to hunt": to go in pursuit of wild animals or game; to engage in the chase; also of animals: to pursue their prey. I can imagine an objection along these lines, however: might we be said to hunt a whole population that is claimed to be invasive? This is not the case, however, for we are not doing that for food or sport, as the definition would require. Rather, we are said to attempt to eradicate or destroy or exterminate a population, just as KSA 80–1202 would have it.

Next, one does not "pursue" a population, for again the target of a hunt is not a whole population. Why? According to the OED, "to pursue" in its transitive form is (originally) to follow (a person, animal, or thing) with intent to overtake and capture, harm, or kill; to hunt; later usually more generally: to chase, go after. And in its intransitive form, "to pursue" is to go in chase or pursuit; to give chase.

Does one "shoot" a population? Again, the answer is no, for one does not shoot a population for the same reasons as one doesn't hunt a population. "To shoot" (intransitive) means the following: of an inanimate thing (or of a living being moving involuntarily): to go or pass with a sudden swift movement through space; to rush, be precipitated; to fly as an arrow from a bow. The point here is this: try targeting an entire population with an arrow!

We need not even consider the definitions of "wound" and "kill," for given background biological knowledge, we can safely rule out these terms being used to apply to whole populations. One cannot wound a population, for to wound implies tissue damage, and populations aren't made of tissues; rather, organisms are. And one cannot kill a population, for a population itself is not alive. (Of course, populations are made up of living things, as are species, but just as it would be literally nonsensical to say that a species can be killed—what we say is that species become extinct—it would be nonsensical to say that we can kill a population.)[33]

Do we trap populations? Again, the answer is no, for traps aren't made to ensnare whole populations. "To trap" in its transitive sense is to catch in or as in a trap, entrap, ensnare. "To trap" in its intransitive sense is to practice catching wild animals in traps for their furs; also generally to set traps for game.

Can we be said to capture populations? This is perhaps a more plausible way of speaking, but again, the verdict is that we don't capture populations. The only place that I found the term being used in any remotely plausible sense for populations is when I discovered the capture-recapture method of population-size measurement that field population biologists use. But what happens in capture-re-capture is that a portion of a population is captured and then recaptured in order to measure the population size. In other words, that portion is an individual or individuals, not the whole population.

Finally, we logically cannot collect a population, for it is nonsensical to collect a collection itself; rather, we collect individuals of a population. To understand why, "to collect" means to gather together *into* one place or group, to gather, get together.

Objections to Scalia's Argument

I will now consider objections to Scalia's argument, one of which will take us back to §1538(a)(1)(B): "It is unlawful for any person subject to the jurisdiction of the United States to . . . take any such species within the United States or the territorial sea of the United States." Once again, our knowledge of taxonomic levels will be very helpful here in understanding the objection and in understanding how one defending Scalia (such as I) might respond to the objection. §1538(a)(1)(B) cannot refer to just whole species, as the text seems literally to read. Scalia gives an excellent argument for this, one that relies on how the ESA is used in prosecuting takes.[34] One can be prosecuted if one takes a single bald eagle, and the relevant part of the statute that could be used in such a prosecution would be §1538(a)(1)(B). Assuming this prosecutorial action is legally permissible, then §1538(a)(1)(B) cannot refer to just whole species, for when one takes a single bald eagle, one is not engaging in the take of the bald eagle species.[35] As this objection continues, §1538(a)(1)(B) could refer, in the logically and physically possible sense of "could," to individuals, populations, and even whole species, on the supposition that the whole species is the last remaining population(s) of the species. In other words, we can *imagine* takes of populations and whole species in our world (logical possibility) in the way that our world is *constituted* (physical possibility).

Given this expansiveness of what §1538(a)(1)(B) is referring to, one could argue that Scalia is wrong to suggest that it only refers to individual organisms. And given this expansiveness, then §1532(19) (the "take" definition of the ESA) could apply to individuals and populations, supposing that we should understand "take" in a consistent sense across the whole of the ESA. And given this expansiveness, then even if breeding in the FWS's definition of "harm" only applies to populations, then Babbitt's regulation (or the FWS definition) is apropos, for Babbitt's regulation would be referring to individuals *and* populations.

How do we respond to this objection, if we find Scalia's original reasoning and my expansion of his reasoning persuasive? We must respond by ignoring the philosopher's hope of the logically and physically possible use of "could" in the above point about §1538(a)(1)(B) possibly referring to taxonomic levels above the organism. Indeed, I admit that it is logically and physically possible to take a population or even a whole population. But that is according to a different definition of "take" than what we have in the ESA. As argued above, "take" in §1532(19) applies to individual organisms, not populations or species. And for §1538(a)(1)(B) to be consistent with §1532(19), then "take" in §1538(a)(1)(B) must legally refer only to organisms rather than populations or species. The philosopher's hopes of expanding takes are dashed once the philosopher is taken down from the clouds of logical and physical possibility and back to the text of the law.

Also, independently of depending on Scalia's statutory method of semantic textualism, there is evidence from a congressional report that §1532(19) refers to organisms rather than populations or organisms. Scalia would not ask that we investigate the intentions of the authors of statutes when we are interpreting statutes, but at this point in the dialectic between the objector and me, it would be useful to appeal to independent support of the view that §1532(19) refers only to organisms. In the Report of the Senate Committee on Commerce of the Senate's version of the ESA, then chair Warren Magnuson wrote, "'Take' is defined in section [§1532(19)] in the broadest possible manner to include every conceivable way in which a person can 'take' or attempt to 'take' any fish or wildlife."[36] And as we saw earlier from §1532(8), "the term 'fish or wildlife' means any *member* of the animal kingdom," that is, any animal *organism*.

The other objection is from Justice O'Connor, who sides with the majority for the most part in reasoning to the conclusion that Babbitt did not overstep his regulatory power in promulgating 50 CFR 17.3(c), but who provides a separate opinion because her opinion is different in some respects. Those differences need not concern us here, but what I will focus upon is her criticism of Scalia's claim that impairment of breeding behavior cannot harm individual organisms. She first uses a thought experiment to imagine the last piece of habitat for the piping plover being destroyed, thereby making it impossible for the piping plover to reproduce. She acknowledges that the last remaining population would be injured, as the species' extinction would loom on the generational horizon, but also argues that the individual plovers would be injured, in the same way that if we were to sterilize a plover, that would injure the bird. To injure is to impair, and to make it impossible for an individual organism to reproduce is to "impair its most essential physical functions and render that animal, and its genetic material, biologically obsolete. This, in my view, is actual injury."[37] O'Connor then concedes that even if impeding an individual's ability to breed is not intrinsically an injury to that animal, interference with interbreeding can cause an animal to suffer obvious physical injury in certain cases. For example, suppose an animal's habitat is destroyed such that the animal cannot flee the habitat to find a suitable new one because of its vulnerable breeding state, or suppose pollutants cause an animal to suffer complications during gestation; in these kinds of cases, physical harm comes to individual organisms partly because they are in a breeding state.

What are we to make of O'Connor's reasoning? The first thing to note is that her last point is something that Scalia would agree with: obviously if certain human behavior causes physical harm to an organism partly because of that organism going through the reproduction process at that time, then that would count as clear harm to the individual. But what we are concerned with here is whether *any* impairment of an organism's ability to breed is an injury to the organism; that is, we are considering O'Connor's concern about whether impairing an organism's ability to breed is *in itself* an injury to that animal. Her objection to Scalia gets to the heart of what I am most interested in here, which is whether it makes sense to say that we can harm populations or species understood as meta-populations, whether impairing breeding behavior applies to the taxonomic level of the population (or even a species) or whether it can apply to individuals who make up the species, and relatedly, whether, on the other side of things, reproduction can actually be *bad* for individuals. This last issue isn't brought up between Scalia and O'Connor, but an investigation of it will help with our analysis.

Taking the following from O'Connor's dissent, that to make it impossible for an organism to reproduce is to "impair its most essential physical functions and render that animal, and its genetic material, biologically obsolete," there is a philosophical view that underlies it, the contemporary natural goodness account.[38] This account began to be developed by Michael Thompson and Philippa Foot, and was most recently articulated by Judith Crane and Ronald Sandler.[39] It is an account that features prominently in debates in ethical theory more generally and in environmental ethics/philosophy more specifically. Let us begin an explanation of this view by considering what is termed the Aristotelian categorical. The Aristotelian categorical is in the form of "Ss are Fs," where S is a variable for a species or a higher taxon and F is a variable for a predicate that provides information about the distinctive life form of individual organisms in that taxon. Examples are "European rabbits are herbivores," "cats have four legs," "the deer is an animal whose form of defense is flight," "the sea turtle makes it alive to the open ocean once hatched," and "after a life in the ocean, a Pacific salmon returns to the freshwater source of its origin and reproduces." Not just any categorical of the form, "Ss are Fs" will count as an Aristotelian categorical. For example, "the blue tit has a round patch on its head" will not count as such.

So, just what counts as an Aristotelian categorical, and how are we supposed to interpret such categoricals on the natural goodness account? With each of these examples, the Fs are *teleological*, that is, they are directed toward an end or towards ends. Such ends include self-maintenance (or survival), reproduction, and sociability. This is why the categorical "the blue tit has a round patch on its head" is not an Aristotelian categorical; it is not a characteristic, as far as we know, that plays a role in the blue tit's self-maintenance, reproduction, and sociability. And I say that this is the account that underlies O'Connor's dissent, because in this account, just as in her dissent, the function or end of reproduction plays a central role in the life of the organism.[40]

Crane and Sandler say that implicit in this analysis of Aristotelian categoricals, there is a species concept, the Axiological Species Concept, at work: "An axiospecies (or axiogrouping [as in a higher-order taxon like a genus, for example]) is a biologically related group of organisms that shares a life form, as described by a set of Aristotelian categoricals."[41] So, for example, in the Aristotelian categorical "European rabbits are herbivores," the species is the European rabbit (*Oryctolagus cuniculus*), which is composed of organisms that share a life form, as described by such categoricals as "European rabbits are herbivores."

There are many problems with the ASC, however, and unraveling these problems will thereby show that the contemporary natural goodness account is problematic, and so then show that O'Connor's reasoning is also problematic. First, the ASC fails to provide a definition of a *species*, which is what species concepts attempt to do. Let us return to what we learned at the beginning of this chapter: one biological question about species is in virtue of what are species distinct from higher-order biological taxa. On the Linnaean hierarchy, this is to inquire about the nature of the species category (or rank) as a taxonomic category, and how it differs from the genus category (or rank), for example.

The ASC straightforwardly does not distinguish species from higher-order Linnaean taxa like genera, or higher-order phylogenetic taxa, as in clades. Crane and Sandler admit as much.[42] It is surprising that they do not view this as a problem, but it clearly is a problem, for what is the point of a purported species concept that does not distinguish species from other taxonomic categories? Why call it a *species* concept, for that matter, at all? Why not just call it an axiogrouping concept?

Independently of this issue, there is a much more serious problem with their concept, a problem that can be viewed as a decisive objection to their concept. As I have told my philosophy students in the past, there are always objections to philosophical views, but only some are decisive, and when a decisive objection can be clearly identified, then that is reason enough to reject the view. One implication of the ASC is that if the Aristotelian categoricals for an axiospecies change, then speciation is occurring. For example, if some species stopped using a certain ecological niche and then switched to a new one, where using the first niche was part of what those organisms that belonged to the species needed to do to survive at first, and then where using the second niche was part of what those organisms needed to do to survive next, then Crane and Sandler are forced to acknowledge that a new species has formed. Using particular niches as what species need to do to survive is part of those species' distinctive life forms, in their view. As a logical implication, they therefore cannot even be open to the claim that the same species had merely evolved. But now this concept is incompatible with the clear empirical fact that species evolve, and can evolve without thereby becoming new species. In fact, though they recognize that a species concept must be explanatorily useful, they fail to recognize that it also must be *evolutionarily* explanatorily useful. If they had realized this, then they might have also recognized that the ASC is defunct, because it is *not* evolutionarily explanatorily useful.

This is an even more significant problem for Sandler and Crane. In another piece, they discuss the Ecological Species Concept and Lawrence Johnson's appeal to something like that concept. This concept takes seriously the idea that ecosystems or environments that species inhabit are central in defining species. In reference to Johnson's appeal, they note:

> Johnson may have in mind here that species often persist within certain parameters, filling certain roles in ecosystems. . . . But a species may also persist when it undergoes significant adaptations such that it does not maintain itself within a fixed range of states understood in terms of ecosystemic role, since species sometimes overrun or transform the ecosystem of which they are part. To suppose that species . . . maintain themselves within these parameters, is to suppose a *static species concept* of the sort that evolutionary theory makes untenable.[43]

So, a species could overrun or transform the ecosystem of which it is a part and yet remain the same species, according to their view here. In other words, in overrunning or transforming the ecosystem that the species inhabited, it is changing ecological niches that it uses. But, if some species stopped using a certain ecological niche and then switched to a new one, then Crane and Sandler are forced to acknowledge that a new species has formed. At worst, this is outright contradiction, but, at best, there is significant tension here—tension that requires resolution (if such resolution is possible, which is doubtful). In other words, it appears that Crane and Sandler have just proposed a static species concept, which is, in their own words, "of the sort that evolutionary theory makes untenable."

To review, then, we found that what was underlying O'Connor's view was the contemporary natural goodness account, and we saw that this account relies on the Axiological Species Concept. Because a decisive objection has been given against the ASC, then at the very least, the contemporary natural goodness account is problematic and so is O'Connor's reasoning.

To show further problems with O'Connor's analysis, there are organisms for which reproduction *can* be bad for them and there are other organisms for which reproduction *is* positively bad for them.[44] The following analysis, therefore, is engaged to undermine the assumption that breeding is always good for the individual organisms that engage in it. Taking the former point first, the birds of the near-threatened Laysan Albatross (*Phoebastria immutablilis*), Black-footed Albatross (*Phoebastria nigripes*), and threatened Wandering Albatross (*Diomedea exulans*) provide superb illustrations of the immense parental investment in rearing young. To an extent not achieved by other birds, albatross birds spend most of their life in the air; in fact, these birds have been known to cover several hundred miles of ocean in one day. But despite spending much of their time "wandering" along the open seas, such birds spend considerable time and energy rearing their young. During a breeding season, a single egg is laid and then each parent takes turns incubating the egg while the other flies hundreds of miles in search of food. As the one parent is incubating and waiting for food, that parent is unable to eat on its own, which often leads to malnourishment, since the one waiting is in stasis for periods lasting sometimes twenty or more days at a time. Once the chick is hatched, the chick must be carefully brooded. Once the chick is large enough to be left on its own (for a while, anyway), both parents head to sea in search of the huge quantities of food that it takes to satisfy a growing chick. Only when the chick reaches approximately the size of its parents is it ready to fledge.

The parental investment in rearing the young implies that would-be parents do not simply start reproducing at the age when they reach sexual maturity. It takes most birds a few more years after that point to begin the breeding process, only after they have gained the knowledge necessary to be able to rear a chick, and only after they have engaged in a long, elaborate courting process to find a suitable mate (they typically mate for life) that they can trust to engage in the intensive time- and energy-consuming process that will ensue.

This story of the albatross birds helps us to see how what is considered typical breeding behavior *can* be bad for individual organisms. The father of environmental ethics, Holmes Rolston, has something particularly apropos to add to our discussion: "Reproduction is typically assumed to be a need of individuals, but since any particular individual can flourish somatically without reproducing at all, *indeed may be put through duress and risk or spend much energy reproducing*, by another logic we can interpret reproduction as the species keeping up its own kind by reenacting itself again and again, individual after individual."[45]

Furthermore, John Nolt reminds us of cases where reproduction *is* positively bad for an individual's flourishing or well-being.[46] In some species, females perish after laying their eggs. In some species of insects and spiders, the female devours the male after mating. And in other species—for example, in species of the aforementioned Pacific salmon—reproduction through spawning causes both the male and female to perish.

Nolt is implicitly referring to the distinction between semelparity and iteroparity in some of his examples. Semelparous species are like those of the Pacific salmon species, in which after a life in the ocean, a Pacific salmon returns to the freshwater source of its origin, reproduces, and then dies. Other semelparous species include many insect species and some bamboo species. Organisms of semelparous species have "life histories characterized by death after first reproduction."[47] The alternative to semelparity, which is living to reproduce repeatedly, is termed iteroparity. It seems very difficult to see how reproduction could be good for the *individual organisms* of semelparous species.

If reproduction is bad for organisms of certain species, then it cannot logically be the case that an impairment of these organisms' ability to breed is an injury to these organisms, for in the case of organisms of semelparous species, we would actually *benefit* the individual of the species if we were to interfere with its breeding process. In other words, given that we could benefit the individual Pacific salmon by interfering with its breeding behavior prior to spawning time, then impairing an organism's ability to breed cannot be an injury to that salmon. Of course, it would negatively impact the population of Pacific salmon in question if we kept the salmon from spawning, but the individual salmon would live a longer life, *ceteris paribus*, by preventing it from spawning.

The ESA and Breeding Behavior

Where does this analysis of breeding behavior lead us? What does it mean for how the ESA is interpreted, or how it should be interpreted? Should it be amended? Or should 50 CFR 17.3(c) be amended? Or should both be amended? Our chapter began with a species with roots in the American Midwest, the black-footed ferret, and so it will close with another species that has an important history in the American Midwest, the monarch butterfly (*Danaus plexippus*). The primary migration corridor of the butterfly is found in this region, as well as its primary breeding habitat. In fact, the USDA's Natural Resources Conservation Service (NRCS) has targeted Midwestern and southern Great Plains states (Illinois, Indiana, Iowa, Kansas, Minnesota, Missouri, Ohio, Oklahoma, Texas, and Wisconsin) in its program to combat the decline of the monarch by planting milkweed and other plants that the monarch feed upon.[48] In its caterpillar/larval stage, only milkweed serves as the source of food, whereas adult monarchs can feed on many different plants. And it's only in milkweed that eggs are laid.[49] In other words, the monarch butterfly needs a certain habitat for breeding, as in milkweed habitat, that is not necessary for non-breeding behaviors, as the adult monarch can feed on many different plants. Eliminating milkweed habitat does not affect the feeding or sheltering behaviors of individual adult monarchs in any meaningful sense, but it does clearly affect the ability of monarchs to reproduce, for if there are no milkweed plants for female adult monarchs to lay their eggs upon and if butterflies in their caterpillar stage do not have milkweed to feed upon, then the monarch species will soon come to an end.

What our analysis and consideration of species like the monarch show is that the ESA or the harm regulation should be amended to prevent "breeding" from being read as an unlawful addition, as Scalia reads Babbitt's harm regulation. What I first suggest is that 50 CFR 17.3(c) be amended to eliminate "breeding" being part of the definition, so that it now reads thusly: "significant habitat modification or degradation where it actually kills or injures wildlife by significantly impairing essential behavioral patterns, including ~~breeding,~~ feeding, or sheltering." This amendment makes it clear that we can only harm individual organisms rather than also populations, with respect to the ESA and to the Code that facilitates the ESA's understanding and execution. (Notably, this also is consistent with much of the environmental ethics literature that maintains that entities above the individual organism in our taxonomic hierarchy cannot literally be harmed.)[50] But then §1538(a)(1)(B) needs to be changed to the following:

> It is unlawful for any person subject to the jurisdiction of the United States to . . . take any such individual organisms of species within the United States or the territorial sea of the United States or to negatively impact a population of said species by impairing the breeding behaviors evident in the population.

This allows "take" to only refer to individual organisms, as §1532(19) and Scalia's interpretation of it would have it; it cleans up the ambiguous language of §1538(a) (1)(B) by inserting "individual organisms of" between "such" and "species"; and it identifies a new category of legally impermissible behavior, that which negatively impacts a population of an endangered species. After all, the purpose of the ESA is to save species, and though it doesn't literally harm or injure a population to erode breeding habitat, it could sensibly be said that in eroding breeding habitat, the population in question is negatively impacted.

So, just as the ESA protects the black-footed ferret from losing its prey species, the prairie dog, so too should the ESA *consistently* protect the monarch from losing its milkweed habitat, that is, if proponents of its being listed as an endangered species under the ESA are ever successful in the listing.[51] Species endemic to the Midwest and species whose members (or parts) spend much of their life in the Midwest need protection by what is one of America's most powerful statutes, the ESA.

Acknowledgments

I want to thank Matt Ferkany, anonymous reviewers, and Burke Griggs for their helpful comments on earlier versions of this chapter.

NOTES

1. Steve Forrest, "Getting the Story Right: A Response to Vermeire and Colleagues," *BioScience* 55, no. 6 (June 2005): 526–530.

2. See Rodrigo Sierra-Corona et al.'s piece for an example of scholarship that criticizes this argument: "Black-Tailed Prairie Dogs, Cattle, and the Conservation of North America's Arid Grasslands," *PLoS ONE* 10, no. 3 (2015): e0118602, https://doi.org/10.1371/journal.pone.0118602. Sierra-Corona et al. argue that the presence of prairie dog colonies can actually benefit cattle. To compare, see Vermeire et al.'s piece "The Prairie Dog Story: Do We Have It Right?" *BioScience* 54, no. 7 (July 2004): 689–695. This is an example of scholarship that partly supports the argument of the ranchers.

3. Sierra-Corona et al., "Black-Tailed Prairie Dogs."

4. There are other laws that are similar to KSA 80–1202 in requiring landowners with prairie dogs to have those dogs exterminated, backed up by penalty of owner's expense in such extermination. And some are in Midwestern states. Such laws include SDCL (South Dakota Codified Laws) 38-22-1.2 and 38-22-17. There is one important difference between these statutes and the Kansan statute, however, which is that in SDCL 38-22-1.2, prairie dogs are conditionally understood as pests, unlike KSA 80–1202, which understands them unconditionally as pests.

5. See Jonathan Proctor, "Larry Haverfield, Wildlife Hero: 1936–2014," *Defenders of Wildlife* (blog), October 1, 2014, https://defenders.org/blog/2014/10/larry-haverfield-wildlife-hero-1936-2014. Interestingly, a statute in another Midwestern state, South Dakota, is specifically about the conditions under which the black-footed ferret can be reintroduced (SDCL 41-11-15).

6. See Miguel Delibes-Mateos et al.'s piece "The Paradox of Keystone Species Persecuted as Pests: A Call for the Conservation of Abundant Small Mammals in Their Native Range," *Biological Conservation* 144, no. 5 (2011): 1335–1346. This is a helpful article on competing ethical and management considerations, including what some might argue is the ethical imperative to successfully bring back the black-footed ferret and the competing goal of certain ranchers to keep the prairie dog from colonizing their properties. In this article, Delibes-Mateos et al. propose a management approach to the prairie dog that is multifaceted in the sense of attempting to adjudicate between these ethical and management considerations.

7. David Jachowski, *Wild Again: The Struggle to Save the Black-Footed Ferret* (Berkeley: University of California Press, 2014), 10.

8. Conrad Hillman and Tim Clark, "Mustela nigripes," *Mammalian Species* 126 (April 1980):

1–3; "Black-Footed Ferret Animal Profile," Black-Footed Ferret Connections, https://blackfootedferret.org/animal-profile/.

9. Technically speaking, the entity is named the Board of County Commissioners of Logan County.

10. Endangered Species Act of 1973, 16 U.S.C. § 1531–1544 (2004).

11. U.S. Const., art. VI, cl. 2.

12. Barnhardt v. Board of County Commissioners of Logan County, 281 P.3d 179 (Kansas Court of Appeals 2012).

13. Note to the reader: this and the following two statutes/regulations are in bold, as I will return to them frequently later.

14. David Adams, ed., *Philosophical Problems in the Law*, 5th ed. (Boston: Cengage Learning, 2013), 202.

15. Babbitt v. Sweet Home Chapter of Communities for a Great Oregon, 515 U.S. 687 (1995).

16. Ian A. Smith, *The Intrinsic Value of Endangered Species* (New York: Routledge Press, 2016).

17. For a thorough discussion of what the metaphysical question is, see Smith's *Intrinsic Value*, 20–21.

18. Whether groups are classified or systematized depends on whether the groups form classes or systems. Linnaeus believed that the groups formed logical classes whereas cladists believe that the groups form systems of nested wholes.

19. The previous two paragraphs are adapted from Smith, *Intrinsic Value*, 20–21.

20. Whether they are parts or members depends on whether species are metaphysically understood to be wholes or sets. If species are understood as wholes, then individual organisms are parts, and if they are understood as sets then individual organisms are members. See Smith's *Intrinsic Value*, 20–21, for further elaboration.

21. Antonin Scalia and Bryan Garner, *Reading Law: The Interpretation of Legal Texts* (St. Paul, MN: Thomson/West, 2012), 167.

22. Babbitt v. Sweet Home, 515 U.S. at 694; Scalia and Garner, *Reading Law*, 168, 195–198.

23. Beecham v. United States, 511 U.S. 368 (1994), at 371.

24. "Alan Dershowitz Defense Argument Transcript: Trump Impeachment Trial January 27," *Rev Transcripts* (blog), January 28, 2020, https://www.rev.com/blog/transcripts/alan-dershowitz-defense-argument-transcript-trump-impeachment-trial-january-27.

25. U.S. Const., art. II, § 4.

26. Nothing in the above summary of Curtis's argument should be taken to imply that I accept his argument; rather, I am just rehearsing it for illustrative purposes.

27. There is a disanalogy between *noscitur a sociis* as applied to "Mantle . . . and other great

competitors" and "treason, bribery, and other high crimes and misdemeanors" on the one hand and the ESA definition of "take" on the other, as the ESA definition does not have an open-ended "and other X" conjunct. However, this disanalogy is not relevant to Scalia's argument. In two cases there is an "and other X" conjunct whereas in the third there is not, but the fact that there are features of commonality that provide the same type of logical function in each case is why *noscitur a sociis* can be applied to all of the cases.

28. In terms that the logic or linguistics student may be familiar with, the OED (https://www.oed.com) provides the very best lexical definitions of the terms in the English language.

29. A classic strict constructionist reading of a Supreme Court case is that of the majority opinion in *Smith v. U.S.*, 508 U.S. 223 (1993). For accessible summaries of (most) Supreme Court cases, see Oyez (https://www.oyez.org) and then search for the SCOTUS case in question.

30. See Ronald Dworkin's "Comment on Scalia," in *Philosophical Problems in the Law*, 5th ed., ed. David Adams (Boston: Cengage Learning, 2013). Ronald Dworkin understands Justice Scalia as a semantic textualist rather than an expectation textualist. A helpful reading of a Supreme Court case in understanding Scalia's semantic textualism is his dissenting or minority opinion in *Smith v. U.S.*, 508 U.S. 223 (1993).

31. A classic expectation textualist reading of a Supreme Court case is that of the majority opinion in *Church of the Holy Trinity v. U.S.*, 143 U.S. (1892).

32. In what follows, I will just be reviewing the relevant transitive and intransitive verb forms of the remaining terms.

33. See Holmes Rolston's "Why Species Matter," in *The Environmental Ethics and Policy Book*, 3rd ed., ed. Donald VanDeVeer and Christine Pierce (Belmont, CA: Thomson/Wadsworth, 2003). Rolston does refer to the extinction of a species as a "super killing," but his language cannot be taken as anything other than metaphorically or figuratively, as species are not themselves alive. And see Smith's *Intrinsic Value*, 103–104, for further discussion.

34. Babbitt v. Sweet Home, 515 U.S. at 719.

35. I realize that "engaging in the take" is a cumbersome locution, but "takings" refer to when the government obtains one's property through eminent domain, physical occupation, or excessive regulation. Thus, I want to avoid confusion here between "takings" and "takes."

36. Senate Comm. on Commerce, Endangered Species Act of 1973: S.1983, S. Rep. No. 93–307, at 7 (1973).

37. Babbitt v. Sweet Home, 515 U.S. at 710.

38. The following seven paragraphs are adapted from Smith, *Intrinsic Value*, 128–131.

39. Michael Thompson, "The Representation of Life," in *Virtues and Reasons*, ed. Rosalind Hursthouse, G. Lawrence, and Warren Quinn (Oxford: Oxford University Press, 1995); Philippa Foot, *Natural Goodness* (Oxford: Clarendon Press, 2001); Judith Crane and Ronald Sandler, "Species Concepts and Natural Goodness," in *Carving Nature at Its Joints: Natural Kinds in Metaphysics and Science*, ed. Joseph Campbell, Michael O'Rourke, and Matthew Slater (Cambridge, MA: MIT Press, 2011).

40. Now, even though a majority of Ss may not, in fact, be Fs, *all* Ss will strive towards such ends as self-maintenance, reproduction, and sociability, in Crane and Sandler's view. This is where the sea-turtle example becomes particularly relevant: even though only a small minority of sea turtles make it alive to the open ocean once hatched, a majority of them try to make it alive to the open ocean because all of them try to survive. And when an individual sea turtle does make it alive to the open ocean and so thereby survives, we will make an evaluation to the effect that the sea turtle's goods have been realized. But when a tiger in the wild, for example, is born without all four legs and thereby suffers and eventually passes as a result, we will make an evaluation that the tiger's goods have not been realized. In short, we will say that this tiger is defective.

41. Crane and Sandler, "Species Concepts," 300.

42. Crane and Sandler, "Species Concepts," 302.

43. Ronald Sandler and Judith Crane, "On the Moral Considerability of *Homo sapiens* and Other Species," *Environmental Values* 15 (2006): 69–84, at 73.

44. The following five paragraphs are adapted from Smith, *Intrinsic Value*, 90–91. And the following examples about albatross birds are discussed in Thom Van Dooren's *Flight Ways: Life and Loss at the Edge of Extinction* (New York: Columbia University Press, 2014), 21–29.

45. Holmes Rolston, *Environmental Ethics: Duties to and Values in the Natural World* (Philadelphia: Temple University Press, 1988), 148, italics mine.

46. John Nolt, *Environmental Ethics for the Long Term* (New York: Routledge Press, 2015), 175–176.

47. Truman Young, "Semelparity and Iteroparity," *Nature Education Knowledge* 3, no. 10 (2010): 2.

48. "Monarch Butterflies," USDA Natural Resources Conservation Service, https://www.nrcs.usda.gov/wps/portal/nrcs/detail/national/plantsanimals/pollinate/?cid=nrcseprd402207.

49. Matthew Faldyn, Mark Hunter, and Bret Eldred, "Climate Change and an Invasive,

Tropical Milkweed: An Ecological Trap for Monarch Butterflies," *Ecology* 99, no. 5 (2018): 1031–1038.

50. For example, see Paul Taylor, *Respect for Nature* (Princeton, NJ: Princeton University Press, 1986).

51. See "Monarch Butterfly," U.S. Fish and Wildlife Service, updated December 15, 2020, https://www.fws.gov/savethemonarch/. The FWS determined on December 15, 2020, that "listing the monarch under the ESA is warranted but precluded at this time by higher priority listing actions. With this finding, the monarch butterfly becomes a candidate for listing; we will review its status each year until we are able to begin developing a proposal to list the monarch."

SUGGESTED READINGS

Adams, David, ed. *Philosophical Problems in the Law*. 5th ed. Boston: Cengage Learning, 2013.

Babbitt v. Sweet Home Chapter of Communities for a Great Oregon, 515 U.S. 687 (1995).

Crane, Judith, and Ronald Sandler. "Species Concepts and Natural Goodness." In *Carving Nature at Its Joints: Natural Kinds in Metaphysics and Science*, edited by Joseph Campbell, Michael O'Rourke, and Matthew Slater, 289–311. Cambridge, MA: MIT Press, 2011.

Delibes-Mateos, Miguel, et al. "The Paradox of Keystone Species Persecuted as Pests: A Call for the Conservation of Abundant Small Mammals in Their Native Range." *Biological Conservation* 144, no. 5 (2011): 1335–1346.

Dworkin, Ronald. "Comment on Scalia." In *Philosophical Problems in the Law*, 5th ed., edited by David Adams, 219–224. Boston: Cengage Learning, 2013.

Foot, Philippa. *Natural Goodness*. Oxford: Clarendon Press, 2001.

Jachowski, David. *Wild Again: The Struggle to Save the Black-Footed Ferret*. Berkeley: University of California Press, 2014.

Scalia, Antonin, and Bryan Garner. *Reading Law: The Interpretation of Legal Texts*. St. Paul, MN: Thomson/West, 2012.

Sierra-Corona, Rodrigo, et al. "Black-Tailed Prairie Dogs, Cattle, and the Conservation of North America's Arid Grasslands." *PLoS ONE* 10, no. 3 (2015): e0118602. https://doi.org/10.1371/journal.pone.0118602.

Smith, Ian A. *The Intrinsic Value of Endangered Species*. New York: Routledge Press, 2016.

Thompson, Michael. "The Representation of Life." In *Virtues and Reasons*, edited by Rosalind Hursthouse, G. Lawrence, and Warren Quinn, 247–296. Oxford: Oxford University Press, 1995.

Vermeire, Lance T., et al. "The Prairie Dog Story: Do We Have It Right?" *BioScience* 54, no. 7 (July 2004): 689–695.

Another Dam Controversy

The Case of the Cuyahoga River from Stagnant
Sludge to Poster Child of the Clean Water Act

Joel MacClellan

The Cuyahoga River is a small river in Northeast Ohio with an outsized influence in U.S. environmental history. The river became heavily polluted in the late nineteenth and twentieth centuries, catching fire several times. The 1969 river fire ignited the public imagination, galvanized the environmental movement, and spurred the creation of the federal Environmental Protection Agency (EPA) and Clean Water Act. Water quality has since improved markedly, yet several controversial dams continue to obstruct the Cuyahoga's flow, reducing environmental quality. The U.S. and Ohio Environmental Protection Agencies recently announced plans to remove all such dams by 2023. There is a compelling environmental history to be told about the Cuyahoga's 100-year transition from a river on fire to a 100-mile long, free flowing river in less than 100 years.

In this chapter, I develop a rich case study and analysis of the Cuyahoga River, rooted in the disciplines and methods of history and philosophy. The history will provide an understanding of shifting uses and attitudes towards America's waters over the course of the twentieth and early twenty-first centuries. The salient history will then be interpreted through the lens of John Dewey's pragmatism.

As the name suggests, pragmatism emphasizes the importance of practical orientation in philosophical thinking, and the unifying tenet of pragmatism is that "knowing the world is inseparable from agency within it"; pragmatists center the interdependency of thought and action, such as the view that knowledge claims be tested by experience or that philosophical speculation absent social progress is not particularly valuable.[1]

The case study will first trace the environmental history of the Cuyahoga in broad brushstrokes, then increase resolution to dam removals on the Cuyahoga, and finally discuss in more detail the controversy, which peaked in the mid-2000s, surrounding the removal of the Gorge Dam—the largest and what will be the last removed dam on the Cuyahoga, which was *simultaneously* being considered for a new hydroelectric power plant and considered for removal to improve water quality. The tension between environmental and recreational values on the one hand and generating hydroelectricity on the other has long been a controversy. Recently, the reinstitution of hydroelectricity generation has been proposed at the site of an existing dam on the Cuyahoga River in Northeast Ohio. This project presents an interesting example because unlike many hydroelectric dam proposals, the dam is already built and has inertia on its side. The environmental stature of the Cuyahoga River adds to the significance of the case.

In what follows, I offer a current history of the state of the Cuyahoga River with an emphasis on the Gorge Dam controversy, argue against the hydroelectric project, and offer a critique of the manner in which the Federal Energy Regulatory Commission (FERC) regulated hydroelectricity in the early 2000s. In the first section I give a brief overview of dams, commonly cited positive and negative aspects of them, and introduce relevant information about regulatory processes concerning dams, particularly FERC, the federal entity which has the authority to license hydroelectric projects. In the next section I present the case of the Cuyahoga, and especially the Gorge Dam case, its context, and the current controversy. In the third section I develop a Deweyian pragmatist approach to understanding the radical shift in the value and treatment of the Cuyahoga River, responding to the material provided in the previous sections. I argue in the next section that this approach is well-suited to make sense of what has happened in human relations with the Cuyahoga River, and that despite taking over twenty years of public negotiation, disapproving hydroelectric dam proposals and removing all dams on the Cuyahoga is the morally right thing to do. I further argue that FERC was maladapted to public involvement at the time and that reform was needed to better involve the public it serves, and offer a sketch of what these reforms might look like.

Two important qualifications about the aim and scope of this essay are necessary before diving in. The first qualification is that the aim is not to pit pragmatism *against* more traditional approaches to moral theorizing and provide a comprehensive and comparative assessment as to why pragmatism is *the* correct moral theory in general or that this approach is uniquely suited to understand the Cuyahoga River case. Such matters (assuming that they are even tractable with argumentation) are far too broad and have nothing to do particularly with the Midwest. The aim is instead to show that a Deweyian pragmatism is *a* fruitful approach for understanding the negotiated shift in values and attitudes concerning our nation's waters that emerged through public dialogue between GOs, NGOs, corporate energy providers, activists, and local residents, not that it is the only approach. Such a Deweyian pragmatist approach offers normative guidance for evaluating public water policy within slow-moving, often reactive public institutions, corporations operating in bad faith with respect to informed dialogue in the public interest, and spirited environmental activists and a lay public trying to have their interests and concerns represented in democratic environmental decision-making processes.

Suffice it to say, however briefly, that pragmatism has certain attractive features as a moral framework for environmental decision-making: it holds a process orientation and an explicitly public conception of the nature of moral discourse as being about problem-solving.[2] This contrasts with more interpersonal moral frameworks such as, say, virtue theory, which emphasizes moral character, or duty-based ethics, which emphasize interpersonal respect, which may not scale well to public solutions and policy choices to environmental problems that are so often collective action problems, not the result of any one individual.[3] Also, the pragmatist approach is explicitly engaged in practical problem-solving rather than arcane philosophical distinctions largely of merely academic interest. For example, this allows sidestepping debate about intrinsic value—the idea that nature has value "in itself" rather than value as a mere resource—instead focusing on Dewey's insights about the role of public reason in science-based public-policy decisions in democratic societies and the tension between technical expertise and democratic decision-making in particular, all of which will be discussed in detail. Science-based public policy includes many important areas, including water management for human and ecosystemic health. On a more conciliatory note, there is much to be gained from multiple perspectives, and not all environmental problems are as public as managing tributaries of the Great Lakes. Pragmatism is surely less valuable in providing guidance as to how I ought to relate to a lone sapling in a forest clearing or how I ought to relate to a member of an endangered species.

The second qualification pertains to Deweyian pragmatism specifically in the broader landscape of environmental pragmatism. The main emphasis here is a Midwestern case study, not a comparative study in historically informed contemporary pragmatisms, but I should speak briefly regarding the broader context. Contemporary environmental pragmatism does glean some from classical American pragmatism, but the emphasis is much less on the details of the historic philosophers than the general method deployed to move beyond philosophical debate and use broad consensus on key environmental topics to motivate policy change.[4] This present inquiry looks back to the history of the Cuyahoga River and a history of the philosophy of John Dewey, but also seeks to apply them to present-day decision-making. Also, it should be noted that Dewey's pragmatism has received relatively less attention than other classical pragmatists, at least in the environmental ethics literature.[5] Perhaps the best-known contemporary defender of pragmatism in environmental ethics is Bryan Norton, who adapts Charles Sanders Peirce's (1839–1914) pragmatism for contemporary environmental concern. What's perhaps most distinctive about Peirce is his pragmatic account of *meaning*: for utterances to be meaningful, they must have practical import. This influence carries over in Norton's emphasis on *communication* and his staunch refusal to affirm or disavow anthropocentrism (a human-centered moral framework), for what can be known and valued in nature, like everything else, must ultimately be explicable in human terms. In Norton's telling, what is most needed is not for there to be more publicly funded science, but rather better communication of what we already know to motivate practical change.[6] There is much to commend about Norton's work, and Dewey was himself influenced by Peirce. What is distinctive about a Deweyian environmental pragmatism is the emphasis on public participation in science and in decision-making about policy, an emphasis on what the public *does* rather than what the public is *told*.

The third qualification concerns environmental justice. In this time of national racial reckoning, there needs to be greater emphasis on linking academic environmental philosophy and environmental justice activists, often led by members of BIPOC communities.[7] Environmental justice is especially important in the Midwest region due to the intersection of the following three facts: the

Midwest is the industrial heartland of the United States, its major cities all have large communities of color, and the burdens of industrial pollution and other externalities all too often disproportionately fall on communities of color. However, these concerns are not paramount in the present study, as the undamming of the Cuyahoga does not nowadays deeply intersect with environmental justice. Most of the course of the upper river is through rural and undeveloped lands. While Cleveland is 40 percent Caucasian and 48.8 percent African American after the Great Migrations of the twentieth century, the Cuyahoga River flows through Industrial Valley and downtown, well outside the African American neighborhoods in East Cleveland and Bellaire–Puritas in the West.[8] The present plan is to store the hazardous dredged sediment from the large Gorge Dam in the adjacent Cascade Valley Metro Park, an area formerly used as an illegal dumping ground. This does not appear to raise any issues of environmental justice.[9]

On the other hand, the early history of settler Americans and Indigenous Americans in the Great Lakes region, inclusive of the area now called the state of Ohio, was rife with injustice. There were forced removals of Indigenous Americans beginning in the late 1700s. Remaining Indigenous American nations banded together to resist resettlement in the Ohio War (1785–1795), the first of many so-called Indian Wars. U.S. territorial sovereignty in Ohio was complete by approximately 1818, whereas construction of the Ohio and Erie Canal occurred in the 1820s and early 1830s, which is to say, that period of colonial brutality was essentially ending just as the story of the Cuyahoga River and its associated canals told below begins.

Context: A Primer on Rivers, Dams, and U.S. Hydroelectric Power Regulation

Rivers often serve as the arteries of civilizations, and many of the world's major cities are at river mouths or at the confluence of rivers. Cleveland, Ohio, is no exception, straddling the mouth of the Cuyahoga River as its waters enter Lake Erie, connecting with the rest of the Great Lakes across the American Midwest, which hold over 20 percent of the world's surface freshwater, and merging with the Atlantic Ocean via the Saint Lawrence River.

Throughout history, human uses of rivers have been manifold. They have served as navigable highways for transportation and shipping, as energy sources to be captured by water wheels and hydroelectric turbines, as fisheries, as recreation for swimmers and boaters, and as sewer ditches for municipal and industrial waste. However, rivers are increasingly appreciated as ecosystems home to environmental values. Due to this new ecological reconceptualization of rivers, dams have come to be viewed in a new light, and environmental impacts are now weighed along with other considerations in public deliberation about dam construction, modification, and removal.

A dam is essentially a plug placed in the path of a river to impede its flow, creating a reservoir by impounding its waters. There are approximately 80,000 large dams (that is, dams over six feet in height) in the United States. However, the golden age of dam construction is now behind us, as the rate of dam decommissioning surpassed the rate of dam construction in the 1990s.

There are several human benefits to damming rivers. Rivers impounded by dams provide a steady supply of water for agricultural, industrial, and domestic uses rather than the boom-and-bust cycle of flooded and low-flow rivers. Hydroelectric power, the vast majority of which is generated by larger dams, is also an important benefit, providing clean energy and constituting part of a clean-energy technologies portfolio to abate climate change.[10] Dams also serve as flood control, protecting low-lying areas susceptible to flooding by controlling water flow and making rivers more navigable for large commercial freighters.[11] Lastly, the reservoirs are often used as recreational areas for boating, fishing, and swimming.

Despite the beneficial aspects of dams just noted, there are also substantial negative consequences. Dams often have high social costs. They can displace local communities by submerging land in reservoirs of water. Also, the larger dams are often built at public expense by the Army Corps of Engineers with most benefits going to private interests, and the public also pays for dam removals once their usable life has passed due to sedimentation and structural decay.[12]

There are environmental costs as well. Rivers are more than just flowing water. They are also partly constituted by the sediments, organisms, etc., found within their waters. River water contains suspended solids such as clay and sand, bodies such as driftwood and other decaying organic material, and living organisms. In the industrial age, rivers also contain elevated amounts of heavy metals and petroleum products. Since dams impede the movement of water, they also impede the movement of sediments and migration of organisms. Reservoir sedimentation greatly reduces sedimentation downriver from a dam, often resulting in delta and estuary erosion. Riverbeds become rockier, resulting in compromised habitat quality for fish and larger invertebrates such as mollusks and crustaceans. Bank scouring results in erosion, making riparian ecosystems vulnerable to invasive plant species. Lower water quality in reservoirs often results as well, including reduced dissolved oxygen levels, high surface temperatures, and cold lower strata, all of which can reduce or eliminate aquatic habitat. Hydroelectric generation dewaters a section of river, diverting the water into tubes and through the turbines, reducing river water and habitat quality.[13]

The Federal Energy Regulatory Commission (FERC), formerly the Federal Power Commission, is an independent agency within the Department of Energy, and was established as part of the Federal Power Act of 1935 to regulate the transmission of natural gas, oil, and electricity. More important to present purposes, FERC is the governmental organization that regulates private hydro-power-generation facilities. Since rivers are a public resource, no private entity can own them. However, privately owned dams on these rivers are permitted. FERC has the exclusive authority to license nonfederal hydropower projects on navigable waterways and federal lands. These renewable licenses are granted for 30–50 year terms. Since the license is a privilege to use public waters, the licensee has the burden of proof to show that public's use.[14]

FERC is a commission of up to five members appointed by the U.S. president and approved by the Senate for a term of five years. The committee must not contain more than three members of any particular political party to prevent undue political pressure and influence.[15] It employs a staff of 1,480 and an annual budget of $347 million. FERC is directly accountable to Congress, yet FERC rulings can also be appealed in federal courts. FERC operates as a quasi-judicial body; it does not manage or set public policy on hydroelectric projects. Rather it decides specific cases, and issues or withholds licenses for hydroelectric projects following a legalistic procedure.[16]

The process that FERC engages begins when a company submits a preliminary application document. FERC then grants a preliminary permit, which is an exclusive right to study the site and file a license during the three-year permit term. FERC coordinates with other relevant governmental agencies, such as the U.S. Department of the Interior and U.S. Environmental Protection Agency, and the licensee in the drafting and execution of a study plan in order to determine the projected social, economic, and environmental costs and benefits of each project. The commissioners vote on each licensing decision (unless uncontested) and either grant or deny the license for a 30–50 year period. Importantly, the public does not play a direct role in FERC's licensing process. The role of the public is limited to a few public hearings where citizens can express their concerns and opinions, and ask questions. Also, citizens can submit evidence in writing throughout the licensing process, which is publicly available via the internet.

The Cuyahoga River: A Natural, Social, and Environmental History

The Cuyahoga River is a small but storied river in Northeast Ohio. The etymology of "Cuyahoga" is unclear, but the Mohawk's "Cayagaga" means "crooked river" and the Seneca's "Cuyohaga" is the "place of the jawbone."[17] The river travels a mere 100 miles down its circuitous path from its headwaters before it dumps into Lake Erie in downtown Cleveland, Ohio, just 35 miles west of where it begins, as shown in Figure 1. Its watershed covers a mere 813 miles. The river's headwaters are remote and pristine, especially considering its proximity to a major city, and after its most dramatic stretch descends through Summit County, the river passes through the heart of Cleveland with high-intensive development along its banks. A damless Cuyahoga would equate to a length of whitewater dropping approximately 200 feet in just 2.5 miles, which is more than Niagara Falls, and would make it one of the longest such stretches of river in the Midwest.

The river was formed during the recession of glaciers at the end of the last ice age, approximately 12,000 years ago, which scoured the earth, creating many of the signature natural features of the Northeast region of the Ohio Valley, and making the Cuyahoga among the youngest rivers on the planet. The ancestors of Indigenous Americans migrated from Asia across Beringia in present-day Alaska and spread across the Americas at about the same time that the Cuyahoga River was forming. The Cuyahoga River and its surrounds became an important area for various Indigenous American peoples and their fishing, hunting, and transportation. In a familiar story across the Northern United States, European settlers and fur traders began arriving in the area and in the sixteenth and seventeenth centuries settled near the river.

The river was briefly a part of the western border of the United States in the late 1790s, making the mouth of the Cuyahoga effectively the northwestern-most corner of the United States for a short period. Moses Cleaveland, a surveyor on expedition throughout the Connecticut Western Reserve, established a settlement

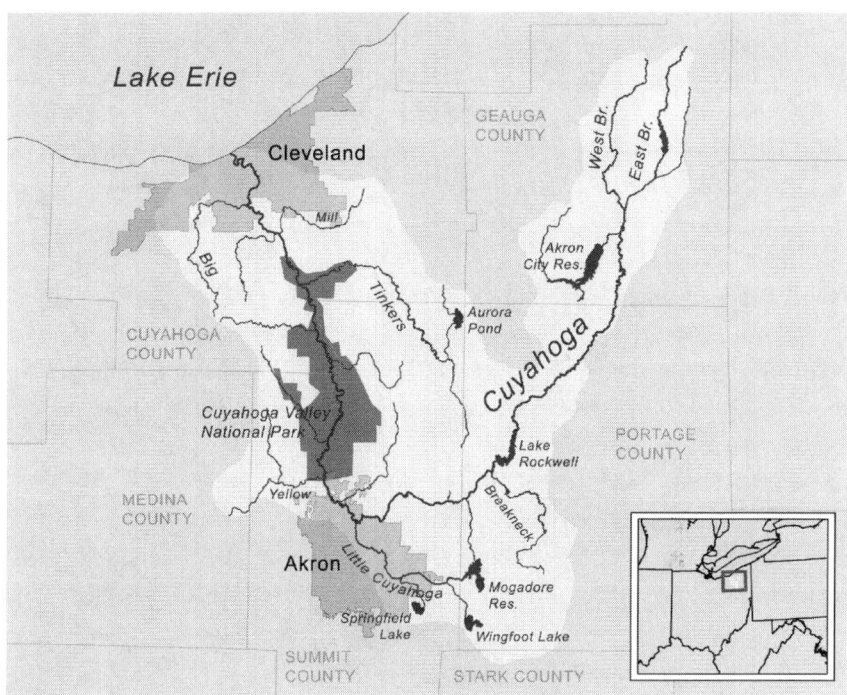

Map of the Cuyahoga River Basin by Karl Musser. CC-BY-SA 3.0
(https://creativecommons.org/licenses/by-sa/3.0/deed.en).

at the mouth of the Cuyahoga River after arriving there on July 22, 1796, which came to be the present-day city of Cleveland. During the brief canal period before the rise of rail, the Ohio and Erie Canal was established. This canal system began near the Cuyahoga River's mouth in Cleveland, carrying freight from 1827 to 1913 from the Great Lakes through Akron and Columbus to the Ohio River, Mississippi River, and ultimately the Gulf of Mexico, opening interior portions of Ohio to regional and national markets. Ohio was the third most economically prosperous state at the height of the canal era in the 1940s, in large part due to the success of its canals. This early success set up the region for rapid industrialization during the nineteenth and early twentieth centuries, with Cleveland becoming a manufacturing hub for steel and sawmills, and Akron, upriver on the Cuyahoga, becoming the "rubber capital of the world" as home to all four major tire-manufacturing companies—General Tire (1869), Goodrich

Corporation (1898), Goodyear Tire and Rubber Company (1898), and Firestone Tire and Rubber Company (1900)—building on the heavy industry infrastructure in the canalway.[18]

The Cuyahoga River became badly polluted, and the factories and industries in and around Cleveland were the chief culprits. Riverside industries dumped their waste into the river with abandon while the Cuyahoga River became heavily polluted as a consequence. The river caught fire at least twelve times during the twentieth century, beginning in 1936—often enough that William Barry, Cleveland fire chief, found the 1969 river fire to be "a run of the mill fire," and it did not merit substantial coverage in the two local newspapers.[19] The 1969 fire was the result of sparks from a passing train igniting oil-soaked floating debris. In the late 1960s, the lower portion of the river, referred to as "the Flats," was effectively an unregulated dump with stretches of the banks lined with abandoned automobiles partially submerged in the river in addition to point-source industrial pollution.

Time magazine published a story about the situation after the 1969 fire, infamously calling the Cuyahoga the river that "oozes rather than flows" in which a person "does not drown but decays."[20] It was largely because of the *Time* article that the national media and national political actors began to focus on the Cuyahoga. The images and accompanying story of a burning river ignited a national environmental consciousness and served as the catalyst for the Clean Water Act (1972) and the Great Lakes Water Quality Agreement (1972). The Clean Water Act was particularly effective at quashing industrial point-source pollution, the primary source of environmental degradation for the Cuyahoga River and connected waterways. Local industries were required by law to clean up their act, economically internalizing their waste, no longer able to free-ride on local human and biotic communities.

The Environmental Protection Agency (EPA) cites the Cuyahoga as essential to the formation of this important regulatory agency: "The 1969 Cuyahoga River fire mobilized public concern across the nation and helped spur an avalanche of water pollution control activities resulting in the Clean Water Act, Great Lakes Water Quality Agreement, and the creation of the federal Environmental Protection Agency and the Ohio Environmental Protection Agency [OEPA]."[21] Much like the Blue Marble, the December 7, 1972, first image of a fragile earth in the void of space taken from the Apollo 17 spacecraft, the image of a burning Cuyahoga became a poster child for the Clean Water Act, an image pictorially representing the often abstract environmental concerns and causes that humans are conceptually and affectively poorly equipped to deal with.

It worked. The environmental quality of the Cuyahoga has improved substantially since the enactment of federal environmental legislation. Under the Great Lakes Water Quality Agreement, the Cuyahoga Remedial Action Plan conducted river cleanup projects at a cost of over $2 billion.[22] A section of the upper river and surrounding valley between Akron and Cleveland was designated a national recreation area in 1974. The river was designated one of the fourteen American Heritage Rivers in 1998. The national recreation area was redesignated as the fifty-sixth national park in 2000.

Cuyahoga Unplugged: The Undamming of an American River

Despite these unqualified successes in cleaning up pollution, portions of the Cuyahoga River remain in noncompliance with the Clean Water Act in the early twenty-first century and this portion of the river remains designated an Area of Concern by OEPA; there remain six literal obstacles to further improve water quality.[23] It is no easy thing to remove a centuries-old dam, and even more remarkable to remove six such dams in a period of approximately twenty years peaking in the year 2005, but this is what has happened, and will happen, to the Cuyahoga after much public deliberation pitting locals against bureaucratic federal institutions. From the headwaters of the Cuyahoga downriver, these are the six dams, with brief histories and their status at the time of writing in February 2020: Kent, Monroe Falls, Sheraton and Powerhouse/LeFever, Brecksville, and Gorge Metro Park Dams.

Kent Dam was built in 1836 to power local mills. With its central location, the dam and reservoir became a part of the city. Kent, home to Kent State University, the site where Ohio National Guardsmen shot unarmed students protesting the U.S. bombing of Cambodia on May 4, 1970, has a strong environmentalist streak as is typical of college towns. However, there were also historical conservationist voices in favor of keeping the nearly two-hundred-year-old structure, which had then become integrated into the local community and had inertia on its side. The old dam was built of hand-cut stone blocks, is the only known stone arch dam with an attached sandstone canal lock, and has been on the National Register of Historic Places since 1977. The land that had been submerged by the dam's reservoir is now a park. That dispute went on for years until a compromise was reached: leave a small portion of the dam as a water park and pool and remove most of the dam to open up the main river channel, allowing river water to run free. This work was completed in 2005 at a cost of just over $5 million.[24] The effect was immediate and dramatic, with the stagnant pool dropping and the river water becoming in compliance with the Clean Water Act to let the river run free. The stagnant water no longer stinks in summer, and Robert Brown, former director of the dam, noticed an immediate uptick in the development and use of Kent's downtown; Cuyahoga Falls mayor Don Walters also saw improvements in the form of kayak and inner-tube rentals and an increase of tourism and travel to the area.[25]

Monroe Falls Dam came down later in 2005. It was originally to be modified rather than removed, but a natural sandstone ledge was discovered during demolition and the dam was removed entirely at a cost of $1.9 million. A natural 4.5-foot waterfall is now in its place.

The Sheraton and Powerhouse/LeFever Dams are smaller dams in Cuyahoga Falls. They were removed in 2013 in a joint venture between Cuyahoga Falls, the Northeast Ohio Regional Sewer District, and the Army Corps of Engineers, returning approximately a mile of river to its natural state.

Brecksville Dam is the last dam along the Cuyahoga, located at river mile 20. The dam was built in 1827 to irrigate the Ohio and Erie Canal with diverted water from the Cuyahoga River. It was retrofitted for industrial use in 1952. It has sat without significant use since 1990. The dam lowers water quality and is a significant obstacle to the movement of fish, waterfowl, and recreational boaters. It has also caused several drownings and is considered dangerous. OEPA has been working with local governments and communities for years to bring it down. The City of Akron has been granted nearly $1 million for its removal, which was slated to happen in 2018 or 2019, and while site preparations have begun, the dam is still intact. This will create a stretch of nearly forty miles of unimpeded river to Lake Erie.[26]

The Gorge Dam is by far the largest dam on the Cuyahoga at 57 feet tall and 420 feet wide. It will be the last dam on the Cuyahoga to come down. It has also been the most hotly contested. The Gorge Dam was built in 1912 near the City of Cuyahoga Falls by the Northern Ohio Traction & Light Company, both as a hydroelectric power plant until 1958 for the streetcar system and to provide water for a coal-fired power plant until the 1990s.[27] It is no small irony that Big Falls, the waterfall for which the city is named, has been submerged by the reservoir for more than a century. The hydropower facility generated approximately 1.5 megawatts (enough electricity for approximately 1,500 homes by today's standards) and was closed in 1958. A coal-burning electricity plant was built nearby, taking advantage of the impounded water to cool its turbines. This facility closed in 1991 and the dam has been functionless ever since.

In 1929, the Northern Ohio Traction & Light Company donated 144 acres to Metro Parks, Serving Summit County, which is now known as Gorge Metro Park. Importantly, the company retained the right to ownership of the dam and the operation of hydroelectric facilities within the park. These rights were transferred to FirstEnergy Corp. (formerly Ohio Edison), the current owners of the dam.

Advanced Hydro Solutions (AHS), a small hydroelectric power company of Fairlawn, Ohio, using its subsidiary company for this project, Metro Hydroelectric Company, submitted a preliminary application document (PAD) to FERC in 2003. This was done to repurpose the Gorge Dam to generate approximately 2.25MW, enough electricity to power around 2,500 homes. FERC approved the PAD in 2004.[28] FirstEnergy supported the proposal, provided AHS with an easement on their rights to generate hydroelectricity in the gorge, and contracted to purchase the electricity generated by the facility, essentially outsourcing the construction and management of the plant while reaping the profits of electricity sale.

While the proposal would have used the existing dam, it would have required the installation of approximately 650 feet of penstock, a 7.5-foot-diameter water-diversion tube beginning at the dam and continuing downriver to the proposed powerhouse location, where the water would power the turbines and be reintroduced into the river. Approximately half a mile of transmission line would need to be installed underground to connect the facility with the nearest electric substation. Also, an access road would need to be constructed, which would entail deforestation of approximately four acres of mature trees on a steep ravine up and out of the gorge. According to a soil survey conducted by the U.S. Department of Agriculture, the soil at the site poses moderate to severe erosion hazards and equipment limitations.[29]

There are both endangered and threatened plant species in the gorge: the endangered northern wood-reed (*Cinna latifolia*) and threatened northern monkshood (*Aconitum noveboracense*). Metro Parks also argued that the project would seriously inhibit their current efforts to repopulate the northern monkshood in Gorge Park by reducing its habitat.[30] The proposed diversion tube and powerhouse would be parallel to the hiking trail on the opposite side of the river, negatively impacting the park's aesthetic value. Gorge Park is well-traveled, receiving approximately 150,000 visitors annually.

The portion of the Cuyahoga River below the Gorge is steep-grade whitewater prized by sport kayakers and canoers. Removing this last dam would make for a 2.5 mile stretch of class III–V whitewater. Recreational kayakers and canoers have long voiced opposition to the dam, as it inundates one of few such stretches in the region, and fully support its removal. OEPA found that the turbulence of the river water's steep decline below the Gorge Dam made it the only section of the Lower Cuyahoga that met minimum water-quality standards, which presumably would not be met due to the diversion of this water through the penstock and turbines.[31] Several rather prominent GOs and NGOs expressed opposition to the project, and a Summit County Planning Commission survey of county elected officials showed that preserving and connecting open spaces and parks was one of the most important local issues.[32] Many impact studies were planned and conducted during and after 2005, which FERC had judged to be relevant to the licensing decision. The final licensing decision from FERC was due in 2009. Already, significant tensions between citizens and local government and GOs on the one hand and federal GOs on the other are clear, a tension that a Deweyian pragmatism developed below helps resolve.

While AHS tried to prolong the life of the Gorge Dam, the Ohio Environmental Protection Agency (OEPA) was simultaneously taking steps to have the dam removed. OEPA issued a report documenting the aspects of noncompliance in the lower Cuyahoga River and what steps would best help to bring that section of the Cuyahoga into compliance with the Clean Water Act, and found that

> Low dissolved oxygen levels, flow alteration and habitat modification associated with reservoir releases and impoundments, and nutrient enrichment were considered the primary causes of impairment [to attaining acceptable water quality]. Further downstream, biological communities improved and marginally met [warm water habitat] biological criteria in the turbulent, free flowing reach between the Ohio Edison dam pool and Little Cuyahoga River. The unique habitat conditions in the gorge help to ameliorate potential water quality impacts from Akron's combined sewers in this section.[33]

In other words, removing the dam would increase the turbidity of a free-running river through the gorge, which would then increase oxygenation. This would improve stagnant conditions for river species and compensate for potentially anoxic conditions caused by algal blooms due to the influx of nutrients caused by fecal matter and detergent phosphates entering the river from sewers during high-rainfall storms. The report also recommended that all dams on the Cuyahoga be removed. OEPA estimated that cleanup could cost as much as $70 million, approximately $57.5 million for sediment removal and disposal and $12.5 million for dam removal and disposal.[34] The costs are so high because of the vast amount of sediment deposited—832,000 cubic yards, or over 90,000 dump trucks worth—over more than one hundred years of interrupting the river's flow.[35] These sediments contain heavy metals, PAHs, PCBs, oil and grease, and pesticides as a result of industrial activity upriver in Cuyahoga Falls and the Akron area over the last one hundred years. There is not significant risk to human health, but releasing several acres worth of contaminated sediment would destroy the river downstream. As has so often been the case in American history, the profits of this dam were privatized, but the environmental and cleanup costs will be externalized to society due to the combined support of OEPA, the federal EPA, local municipalities, and relevant NGOs. Predictably, FirstEnergy was prepared to lease the dam when it came to profiting off the structure, but when it comes to its removal, FirstEnergy has been cagey as to who owns the dam when it comes to footing the bill for its removal. As recently as 2015, Bill Zawiski of OEPA remarked that they did not know who owns the dam.[36]

Ultimately, by all appearances, the people and OEPA won out in this case, ultimately holding federal institutions to account. The hydroelectric project was denied, and the dam will be torn down and the sediment removed jointly by the EPA and OEPA with funding from the Great Lakes Legacy Act and the City of Akron, among other stakeholders, according to Jennifer Conn, reporter for local NPR station WKSU.[37] That project is currently scheduled for 2023, meaning that the process of removing six dams on the Cuyahoga River will have taken approximately twenty-five years since the idea was first put forward to improve water quality and bring the river into full compliance with the Clean Water Act. If all goes according to plan, the Cuyahoga River will flow freely along its entire stretch in 2023 for the first time since 1827.

A Deweyian Pragmatism: A Brief Primer on Pragmatism

The benefit of hindsight of more than twenty years shows that river restoration ultimately won out over hydroelectric power and other concerns. Yet this was far from obvious in the late 1990s and early 2000s. How did river restoration come to win out over clean energy generation and preservation of historic structures? How did local, state/regional, and federal organizations with different mandates resolve their tensions? What roles did the public play in deciding the fate of a storied American river? Was this a wise use of approximately $100 million? In what ways and to what degrees were the decision-making processes democratic, fair, and morally right?

A Deweyian pragmatist framework provides a fruitful moral philosophical framework for satisfactorily responding to the above questions. However, before attending to these questions, some general remarks about Dewey's moral philosophy are needed to provide context.

John Dewey's moral philosophy is difficult to interpret. This is so for several reasons. He did not present his moral philosophy in a unitary magnum opus. Rather, it is scattered across many of his writings. This makes correct interpretation a subtle endeavor and has led to the rather unfair characterizations of Dewey's moral philosophy as a crass instrumentalism, a shill for American industrialism, and therefore a nonstarter for contemporary environmental ethics.[38] Also, Dewey's views differ radically in many ways from much of the Western philosophical canon (as will be argued shortly), and it is a mistake to read Dewey in a modernist light.

Dewey's lukewarm reception among philosophers, while perhaps understandable, is regrettable because there are several aspects of Dewey's work that make it a viable framework for moral reasoning and for environmental problem-solving specifically. These include Dewey's views (1) about the relationship between humans and nature, (2) that a moral theory should focus on deliberative procedure rather than values or norm prescription, (3) that morality is fundamentally social, and (4) about the centrality of empirical inquiry. These will be discussed in turn, as they help the reader understand Dewey's position and are helpful in understanding their application to the Cuyahoga case thereafter.

First, John Dewey is one of the first post-Darwinian moral philosophers. Dewey lacked the firsthand experience of Aldo Leopold, who slaughtered wolves during his tenure as hunter and trapper for the recently established U.S. Forest

Service and later rejuvenated depleted soils in Sand County, nor did Dewey experience sublime rapture as did Arne Naess in the Hindu Kush or John Muir in the High Sierra. Unlike modernist European moral philosophers who developed their theories prior to the publication and dissemination of Charles Darwin's *On the Origin of Species* (1859), such as Immanuel Kant (1724–1804) and John Stuart Mill (1806–1873), John Dewey (1859–1952) developed his theory with evolutionary naturalism as the backdrop, meaning that the natural world is the world (anti-supernaturalism), and that humans' place in that world is as a contingently evolved primate species, not the centerpiece of a divinely designed cosmos. Core to Dewey's philosophy is a critique of dualisms such as man-nature and mind-body, which he saw as entailing subjectivism, the view that truth is determined by each person, which ultimately separates us from others because there is no external or objective truth that connects us and helps adjudicate disagreements. Dewey says subjectivism is both philosophically problematic and unproductive to real-world problem-solving because it denies objective truth and the sociality of problem-solving. Dewey's view of value judgments is naturalistic, based in social psychology and what we would now consider evolutionary and developmental psychology. That Dewey conceived of humans as fundamentally a part of nature and saw philosophy as providing useful guidance makes the view conducive to environmental ethics. This naturalism is a break from Medieval and Enlightenment thinkers such as Augustine, Aquinas, Descartes, and Kant who saw humans as fundamentally separate from and superior to nature. The emphasis on producing a process for moral deliberation rather than arguments for particular moral positions makes pragmatism useful to deal with the real-world problems we face, rather than focusing on making fine-grained distinctions about the nature of intrinsic value or moral considerability as professional philosophers are increasingly wont to do.

A second core aspect of Dewey's moral philosophy typical of American pragmatists but especially pronounced here was his commitment to *process* rather than *product* in moral inquiry. Dewey is not a prescriptivist in that he does not prescribe particular moral values or normative judgments in advance for moral evaluation, but rather endorses a process wherein we discuss and evaluate moral values so as to arrive at moral judgments. He was deeply skeptical of absolutism, the view that moral truths hold for every situation and time, and foundationalism, the theory of knowledge that founds all knowledge on a set of basic truths. Rather, he thought that values were already pervasive in human life; it was not the place of philosophy to dictate what we ought to value. There is a place for principles understood as practically useful encapsulations of our previous experiences. This differs from Kantian deontology for example, which alleges that the categorical imperative, the supreme principle of morality, is discoverable by the intellect alone and differs from rule utilitarianism, which holds that we should follow rules that maximize happiness. As James Gouinlock puts it, Dewey wanted "to discover and clarify the ways in which persons could determine their own fulfillment as natural and social beings. The philosophical task—a formidable one—is to enlighten, not to prescribe, human conduct."[39]

In this way, Dewey's moral philosophy is quite different from most Western moral philosophies, as well as environmental ethical theories such as biocentrism or animal rights theory because such views advocate particular ideas about the nature and extent of moral value, such as the claim that all living things are morally considerable (biocentrism), or the claim that the welfare of sentient beings is of primary importance (animal rights theory). Dewey's moral philosophy isn't meant to provide a set of moral principles that prescribe what we ought to do, such as always respecting life or minimizing rights violations. That said, Dewey's view remains prescriptive in the sense of offering a *method* of arriving at moral prescriptions, rather than in the sense of giving moral prescriptions following from a particular moral principle or pattern of justification.[40]

A related aspect of Dewey's thought and pragmatism generally is the treatment of intrinsic value. This term of art has been the subject of much discussion by environmental philosophers and the subject of a fascinating if not dizzying array of distinctions and terminology.[41] Perhaps the most important such distinction is intrinsic value, understood as noninstrumental value, versus instrumental value, which is value a thing has as means to some end. For example, one can regard a forest as merely instrumentally valuable to producing wood or oxygen, or one can regard a forest as (also) valuable and worth preserving in its own right. Another distinction is between intrinsic and extrinsic value, that is, value that a thing has solely by virtue of its intrinsic properties, and value something has by virtue of its relations to other entities.

Dewey does acknowledge that we talk as if there is intrinsic value, but he notoriously thought that belief in intrinsic value is a mistake. Given the importance of intrinsic value talk in environmental ethics, what should be made of this? One line of response here, which also is the line that Bryan Norton takes, is to accept this conclusion and counter that the richness of instrumental values in nature is greater than is typically supposed, including instrumental uses such as education and not merely economic usefulness, such as seeing a river as merely a navigation corridor, source of water, and pollution sink. To this, it helps to add some context about what Dewey was arguing *against*, which was the traditional view that intrinsic value was supernatural, fixed, and immutable—that is to say, God's commandments are eternal truths. Dewey's worry was that such appeals to intrinsic value stalled human progress, creativity, and empirical inquiry for millennia.[42] On this second strategy, we can understand that aspects of nature may well be intrinsically valuable, not "in themselves" entirely separately from human beings, but as valued in a certain way, namely, valuable not (merely) as a resource for humans (or perhaps humans and other evaluative beings, such as some animals).[43] Suffice it to say that this conception of intrinsic value seems to overcome what Dewey was most concerned about in arguing against intrinsic value, which is really immutable, divinely created, extrinsic value.

Whichever interpretative strategy is best, what is ultimately important is not whether nature is valued according to any particular understanding of instrumental or intrinsic value, but rather the practical import of those convictions. To see this, suppose that a philosopher proves that nature is intrinsically valuable once and for all. *Nothing happens.* It is for this reason that much contemporary environmental pragmatism seeks to sidestep debates about intrinsic value and whether environmental value is anthropocentric or not. For, as a practical matter, it's irrelevant precisely what *kind*(s) of value nature has, and there is an opportunity cost to engaging in such debate, as it's time spent not acting.[44] Suffice it to say that, at least for Dewey, it is far more important that philosophers promote participatory decision-making in environmental policy rather than a pet theory of environmental value.

Third, moral inquiry is a fundamentally social endeavor, according to Dewey. Modern philosophy, including modern moral philosophy, from René Descartes's *cogito ergo sum* (I think, therefore I am) to John Rawls's original position where any person could heuristically consider what would constitute justice for basic social institutions, and where justice does not feature in familial relations but rather as obtaining only amongst strangers in the public sphere, has been highly individualistic. For Dewey, morality is fundamentally about navigating conflict towards resolution and is necessarily a process requiring social deliberation, not careful thought in isolation. Central to morality is what Dewey called "social intelligence," which requires certain norms of engagement with one's peers to resolve conflict, including epistemic norms and values. Democratic institutions are the obvious social framework within which such conflict can be mitigated, but they are not the only ones. Families and other social groups are also locations for the practice of social intelligence.

Lastly, for Dewey, ideas have to be tested for their usefulness to explain, inform, and enrich human experience. Rejecting subjectivism allowed Dewey to maintain that there is no gap between human experience and nature, and that values are functions of natural processes, open to empirical inquiry. In this light, Dewey saw science as a central tool for human understanding of how to best interact with nature. Yet science itself is a fundamentally social endeavor and a particularly reliable exercise of social intelligence. However, Dewey was deeply skeptical about an expert class of scientists or anyone else, always stressing the importance of dialogue between experts and those whom they serve. Dewey cautions against technocracy, or rule by experts, and defends participatory democracy in *The Public and Its Problems* (1927) against skeptics, including Walter Lippmann, who argued for minimally democratic states controlled by technical experts in *Public Opinion* (1922) and *The Phantom Public* (1925).[45] Dewey remained humble about human epistemic limits, including his own, and held a fallibilist epistemology, a theory of knowledge that holds that at least some legitimate knowledge claims are possibly wrong, that is, do not require certainty. The tasks of knowing and problem-solving are always incomplete and open to revision in light of both new information and new concerns.

James Goinlock draws these four threads of Dewey's philosophy together well:

> Actual life activity, as Dewey understands it, is found to be inherently satisfactory when the powers of the individual are effectively integrated with his environment so that the entire inclusive process functions harmoniously. The agencies of both individual and environment are united in an ongoing process; initially diverse or impeded values are unified in a reconstructed situation. . . . Such activity is a unification of values—what Dewey calls consummatory experience. It is a form of organic interaction of man and nature, and it cannot be achieved without a certain amount of knowledge of how human nature and its environment function together. Further, inasmuch as most ventures involve the various activities of a plurality of persons, consummatory experience requires that the persons involved deliberately act in concert. That is, our social condition makes intelligent cooperation a necessary condition for consummatory value.[46]

In short, Dewey's pragmatism, while often misunderstood, has much to recommend it, especially with regard to the kinds of complex problems found in environmental ethics that have multiple parties with conflicting perspectives. Indeed, Dewey's views in many ways anticipate recent philosophical trends inquiring into the proper relationship between science and public policy and social epistemology more broadly, the idea that knowledge is a collective achievement of a community, not an individual intellectual possession as it has often been treated in philosophy.[47]

Pragmatism in Practice: Understanding the Case of the Cuyahoga

With this history of the Cuyahoga River and some exposition of a Deweyian environmental pragmatism now in hand, I will now apply a Deweyian framework to the public problem of what to do about the Cuyahoga's dams, paying special attention to the Gorge Dam and the hydroelectric power plant proposal and FERC's process in dealing with this issue. It will be shown that several aspects of FERC's decision-making process are problematic from a Deweyian perspective, but that this has been improving in recent years in response to changing social circumstances. This is a valuable exercise because it brings to light ways in which the public has navigated the dynamic waters of unprecedented change through rapid industrialization with much unanticipated environmental degradation. The Deweyian analysis shows the importance of the public holding industry accountable. Ultimately, it seems that the democratic process, while perhaps slower and more convoluted than it might have been, embodied social intelligence and has led to a satisfactory conclusion in the case of Cuyahoga River's dams in northeast Ohio.

FERC gave a quite narrow interpretation of the "public interest" language in the Federal Power Act through most of its history by construing hydropower as in the public's interest rather than other concerns, rejecting only one hydropower license application on aesthetic or recreational grounds in its first sixty years of existence. Both Congress and federal courts have authority over FERC, and both changed the basis upon which FERC makes its decisions just before and during the early 2000s when FERC was seriously entertaining the Gorge Dam hydroelectric proposal.

Several judicial developments are relevant here, which suggest that it took quite a bit of work to make FERC more accountable to the public it serves. In *Udall v. Federal Power Commission*, the Supreme Court provided a broader definition of the "public interest" language of the Federal Power Act as it pertains to FERC, reversing FERC's narrow focus on promoting hydropower:

> The grant of authority to the Commission to alienate federal water resources does not, of course, turn simply on whether the project will be beneficial to the licensee. Nor is the test solely whether the region will be able to use the additional power. The test is whether the project will be in the public interest. And the determination can be made only after an exploration of all issues relevant to the "public interest," including power demand and supply, alternate sources of power, the public interest in preserving reaches of wild rivers and wilderness areas, the preservation of fish for commercial and recreational purposes, and the protection of wildlife. The need to destroy the river as a waterway, the desirability of its demise, the choices available to satisfy future demands for energy—these are all relevant to a decision. . . . On our remand there should be an exploration of these neglected phases of the cases, as well as the other points raised by the Secretary."[48]

This language was strengthened by Congress with the Electric Consumers Protection Act of 1986, with the addition of an equal consideration clause:

> In deciding whether to issue any license [for a hydroelectric project], [FERC's] Commission, in addition to the power and development purposes for which licenses are issued *shall give equal consideration* to the purposes of energy conservation, the protection, mitigation of damage to, and enhancement of, fish and wildlife (including related spawning grounds and habitat), the protection of recreational opportunities, and the preservation of other aspects of environmental quality.[49]

There is one other important aspect of the Federal Power Act controlling FERC's licensing process, the "comprehensive plan" language: the commission can approve a plan only on the condition that "The project adopted, including the maps, plans, and specifications shall be such as in the judgment of the Commission will be *best adapted to a comprehensive plan* for improving or developing a waterway . . . for interstate commerce . . . for water power development . . . for adequate protection, mitigation, and enhancement of fish and wildlife, and for other public uses."[50]

The result of these changes in terms of what must be considered when making licensing decisions was well stated by FERC's Timothy Konnert, commenting on the Gorge hydroelectric project proposal:

> In deciding whether to issue a license, the Commission must give equal consideration to developmental and environmental values of the project. Environmental values include fish and wildlife resources, including the spawning grounds and habitat, visual resources, cultural resources, recreational opportunities and other aspects of environmental quality. Developmental values include irrigation, flood control, and water supply. The Commission must also ensure that the project is best adapted for developing the waterway for beneficial public purposes.[51]

The "comprehensive plan" component is especially important here, as there already exists a comprehensive plan for the Cuyahoga River. The Cuyahoga River Area of Concern (formerly the Cuyahoga River Remedial Action Plan) was created in 1988 by OEPA with the mission "to plan and promote the restoration of the environmental quality of the Cuyahoga River through the remediation of existing conditions and prevention of further pollution and other degradation."[52]

A project powering a maximum of 2,500 homes, which would have compromised Gorge Metro Park's mission, posing a threat to endangered and threatened plant species, would have detracted from the recreational experience of 140,000 annual visitors, and would have perpetuated the presence of a dam deemed by OEPA to prevent the attainment of water-quality standards for the Cuyahoga River. It is clear that granting such a license is not in the public interest; the Gorge Dam "promise" should not be perpetuated through a FERC license.

In an interesting turn of events, FERC claimed that it would not require a dam-removal feasibility study as a part of its ruling on the Gorge Dam:

> Because this dam is unlicensed, the Commission has no authority to require its removal. Because the potential application for this project would be for an original license, we do not consider dam removal a reasonable alternative. Alternatives that would be evaluated with a project seeking an original license such as the proposed Metro Project include:
>
> 1. License the project as proposed by the applicant.
> 2. License the project as proposed by the applicant with FERC modifications.

3. A no-action alternative where the Commission does not issue a license and the project is not constructed [but the dam remains].[53]

While FERC does not have the authority to require the dam's removal, this does not mean that the results of such a study are not relevant to deliberation about which path of action best serves the public interest. Since FERC is supposed to give equal concern to environmental values and energy conservation, and since FERC took the dam's presence as a given in their deliberation, FERC seems to have unduly constrained their deliberation's scope, thereby failing to account for the public interest.

However, OEPA rightly noted that FERC *does* have the authority to require the licensee to conduct studies relevant to compliance with other applicable regulations; Section 401 of the Ohio Administrative Code states that a water quality certification is required "for any federal permit or license to conduct any activity which may result in any discharge to waters of the state." Such a certificate will not be granted unless the director determines that the applicant has demonstrated that "any obstruction or alteration in waters of the state will not prevent or interfere with the attainment or maintenance of applicable water quality standards . . . or implementation plans."[54] OEPA had the following to say on the issue:

> Dam removal feasibility studies are one of the recommendations in the US EPA approved Lower Cuyahoga TMDL report. The TMDL is an implementation plan defined in Section 303 of the Clean Water Act. Metro Hydroelectric . . . acknowledges that dam removal is a possible option. Therefore, we believe that a study of the feasibility of dam removal at the Ohio Edison Dam is not beyond the scope of the FERC licensing process.[55]

AHS, the local hydroelectric company, acknowledged that OEPA may order the Gorge Dam's removal even if FERC grants the license, signaling the significant disconnect between federal agencies. The public would have then accrued the additional financial burden of buying AHS out of their license in order to honor its terms, and so as to avoid a lawsuit, to be able to remove the dam. Since FERC was supposed to rule in the public interest and because meeting the water quality conditions set out in the Clean Water Act is in the public interest, it appears problematic that FERC claimed to be ruling in the public interest in accord with a comprehensive plan for the Cuyahoga River by issuing AHS the license.

Furthermore, there is evidence that FERC and AHS were not supportive of much genuine public involvement in the licensing deliberation. Public notice of proposals and solicitation of public comment are required by law, yet according to Friends of the Crooked River, a local NGO, the public notice for the Gorge project appeared in an "obscure weekly paper in southern Summit county," not in the major daily papers in northeast Ohio such as the *Akron Beacon Journal* or *Cleveland Plain Dealer*.[56] If FERC and AHS were truly interested in public involvement, then publication in a minor newspaper is obviously inadequate. Also, since the notice soliciting public comment appeared in mid-March 2004 with the comment period closing on March 30, the public was not given adequate time to consider the issue. Add to this the fact that AHS submitted their PAD to FERC in December 2003, but did not notify Metro Parks until three months later, just before the public comment period closed. All of the above strongly suggest disinterest in genuine public deliberation in the licensing process. If anything, it appears that there were attempts to sidestep the kind of social deliberation that Dewey saw as central to social problem-solving.

Moreover, FERC's quasi-judicial character was itself an impediment to effective public participation in the deliberative process. The rules that govern public participation in FERC have been quite formal in comparison with most other federal agencies. To be recognized in FERC's proceedings, an individual citizen must be an official party to the case, akin to a litigant in court, and in order to become a party in the case, one must file a motion to intervene accompanied by a certificate of service. These legal formalities created high barriers to entry, and these steps were seldom taken except by the most organized, educated, and well-funded parties. While FERC accepted letters in praise and protest of cases as evidence, they are taken far less seriously than statements from parties to the case following the above legalistic formalities.[57] That this is the case is partly shown in the section of FERC's website where all dockets relevant to a particular case are stored; comments received from citizens are listed as "Individual (No Detailed Affiliation Given)," whereas official parties submitting evidence to the project are listed by name.

The problem of inadequate public inclusion in FERC's licensing process was exacerbated during this time. The Interim Final Rules of the new Energy Policy Act of 2005 introduced new criteria slanted towards increasing hydroelectric generation, and they include a very fast timetable and technical burden that cannot be met without significant resources. Furthermore, the new rules themselves took effect without meaningful public comment as to their content. At best, the public could have influenced FERC through the few public meetings that were held and by having submitted evidence in writing, but what occurred is that the public had no decision-making role in FERC's hydropower licensing process. On a more promising note, FERC decided to include a group of kayakers—clear stakeholders in this process—to be involved with scientists studying the impacts of the water diversion on flow levels and the recreational impact of this diversion on the boating community.[58] However, this was a fairly minor component of the overall decision-making process. The more substantial issue is the relatively low degree of public involvement in the deliberative process itself.

The tension between technical expertise and democracy is not altogether new, and there has been an increased call for public involvement in science as well as critical reflection regarding what role values do and should play in the scientific aspects of policy formation in recent years.[59] Dewey was concerned about precisely this issue in *The Public and Its Problems*:

> No government by experts in which the masses do not have the chance to inform the experts as to their needs can be anything but an oligarchy managed in the interests of the few. And the enlightenment must proceed in ways which force the administrative specialists to take account of the needs. . . . The essential need, in other words, is the improvement of the methods and conditions of debate, discussion and persuasion. That is *the* problem of the public.[60]

To this, Dewey adds that the proper role of expertise is in making known the facts upon which judgment depend, not in framing and executing policies. Contrastingly, the public does not need the ability to carry out the investigations, but instead needs the ability to judge the bearing of the knowledge supplied by experts upon common concerns. Dewey sums up the proper relationship between experts and public with his famous shoemaker analogy: "The man who wears the shoe knows best that it pinches and where it pinches, even if the expert shoemaker is the best judge of how the trouble is to be remedied."[61]

With Dewey's words in mind, we have a better understanding of what the proper roles are for expertise and the public. However, the public does not itself have the direct authority to make decisions about hydroelectric project licensing in the United States. David Guston's Principal-Agent model helps us better understand the delegatory relationships between FERC and the public, and the role of expertise in this process. In this model, the principal is an actor who requires a task to be performed but lacks the ability to perform it directly. The agent is an actor whom the principal delegates or contracts to perform the task.[62] In the case at hand, the public is the principal that delegates rule and regulation through voting to Congress, the people's agent. In turn, Congress is the principal, delegating the regulation of hydroelectric projects to its agent, FERC. The commission delegates to its staff and the licensee the task of rounding up the facts and community concerns and values, and it is here where the public has the opportunity to enter the process. While Dewey is right that decision-making should be a matter for the public, this only occurred with a convoluted delegatory relationship in FERC, far from the ideal of a democratic deliberative process.

There seems to be room and desire for a more direct role for public involvement in the process of hydroelectric regulation. In a Deweyian spirit, Heather Douglas proposes the extent to which a process maximizes the interaction between citizens and experts and maximizes the influence they have on one another as a measure for evaluating the role of public participation in technical decision-making. The reason that this is a good criterion is because deliberation is needed to inform analysis just as analysis is needed to inform deliberation, and we can best address our science-based policy questions when experts and citizen-stakeholders work collaboratively.[63] In light of what has been said above, FERC scored low if evaluated on its success on the interaction and influence that citizens and experts have on each other. In the case of the Gorge Dam proposal, public interaction with experts was too limited and took substantial public coordination with other government agencies such as the EPA to free the river.

The U.S. Supreme Court presided over three cases, around that time, of profound importance to the FERC licensing process: *Carabell v. Army Corps of Engineers* (2004), *Rapanos v. United States* (2006), and perhaps most importantly, *S.D. Warren v. Maine Board of Environmental Protection* (2006). In the latter, a landmark case upheld the Maine decision that states may protect rivers against hydroelectric dams and other federally licensed activities under the Clean Water Act. It seems that FERC has gotten the message, as it now does a better job of balancing energy demands and environmental quality, having shifted its orientation in recent years from serving as a rubber-stamp for private hydropower generation towards a more comprehensive view of the public's interest.

Conclusion

The twentieth and early twenty-first centuries saw a remarkable turnaround for the Cuyahoga River from among the world's most polluted rivers to a redemption story in the heart of the Rust Belt of the American Midwest. The American Rivers environmental NGO designated the Cuyahoga the "River of the Year" for 2019 in celebration of the fifty years that have passed since the famous 1969 fires and the river's environmental quality resurgence. The future of the river is brighter than it has been for over one hundred years.[64] A Deweyian pragmatism provides a fruitful lens from which to understand how people navigated emerging problems of water pollution, commerce, clean power generation, and the value of clean waters and ecosystems. It also allows us to see when that deliberative process can go awry, such as when federal regulatory bodies become too detached from the public they serve, as appears to have been the case for at least part of FERC's history of hydroelectric dam regulation. The task of problem-solving is forever incomplete; the dam removals along the Cuyahoga River have led to increased flow, especially after storms, causing increased erosion along parts of its path, which local communities are now endeavoring to resolve or work around. Dewey would have it no other way; the human condition is one of constant problem-solving, but always holds the promise of progress, both for ourselves and in relation to other creatures for whom Earth is home.

NOTES

1. Catherine Legg and Christopher Hookway, "Pragmatism," in *The Stanford Encyclopedia of Philosophy*, ed. Edward Zalta, Spring 2019 ed., https://plato.stanford.edu/entries/pragmatism/. As Legg and Hookway note, this tradition is increasingly recognized as a third alternative to Anglo-American analytic and Continental philosophical traditions.
2. See Andrew Light and Eric Katz, eds., *Environmental Pragmatism* (New York: Routledge, 1996). This is the earliest seminal volume on this topic.
3. Onora O'Neill concedes this point about Kantian ethics, arguing that what it lacks in scope it makes up for in precision for interpersonal ethics. See her "Kantian Approaches to Some Famine Problems," in *Matters of Life and Death*, ed. Tom L.

Beauchamp & Tom Regan (Philadelphia: Temple University Press, 1980), 546–551.

4. See especially the earliest seminal volume on this topic, Light and Katz's *Environmental Pragmatism*, which understandably, given its aim, is much more about using pragmatist methodology to advance the practical use of philosophy in the policy sphere than it is about the historic texts in which they find inspiration.

5. The most notable treatment of Dewey's pragmatism in environmental ethics is Hugh P. McDonald's *John Dewey and Environmental Philosophy* (Albany: State University of New York Press, 2004). One noteworthy critic, C. A. Bowers, attempts to foreclose a Deweyian environmental pragmatism on the grounds that he is essentially a cultural imperialist in his epistemology for advocating for Eurocentric science over indigenous ways of knowing. While Dewey allows for common sense and locally acquired knowledge, it's not clear that his scientific realism requires apology, for in my view science is a more accurate and reliable means of understanding and predicting the world when compared to nonscientific mythical cosmologies. It does bear mention, however, that perhaps due to Dewey's longstanding influence in the philosophy of education, he is an oft-cited inspiration in contemporary work in environmental education. See C. A. Bowers, "The Case against John Dewey as an Environmental and Eco-Justice Philosopher," *Environmental Ethics* 25, no. 1 (2003): 25–42.

6. For Norton's early defense of weak anthropocentrism, see "Environmental Ethics and Weak Anthropocentrism," *Environmental Ethics* 6, no. 2 (1984): 131–148. For the emphasis on communication in public discourse on environmental policy, see *Sustainability: A Philosophy of Adaptive Ecosystem Management* (Chicago: University of Chicago Press, 2005).

7. For a great introduction to environmental justice and its tension with academia, see Kristin Shrader-Frechette's *Environmental Justice: Creating Equity, Reclaiming Democracy* (New York: Oxford University Press, 2002).

8. "Quick Facts: Cleveland," U.S. Census Bureau, https://www.census.gov/quickfacts/clevelandcityohio.

9. "Gorge Dam: Cuyahoga River Area of Concern," Great Lakes Mud, https://www.greatlakesmud.org/gorge-dam—cuyahoga-river-aoc.html.

10. It is worth pointing out that wind power has just eclipsed hydroelectric power as the most-used form of renewable energy in the United States. "Wind Has Surpassed Hydro as Most-Used Renewable Electricity Generation Source in U.S.," U.S. Energy Information Administration, https://www.eia.gov/todayinenergy/detail.php?id=42955#.

11. The benefits of flood control and navigation have increasingly come under attack. Floods vary substantially in magnitude, and since larger dams tend to be more expensive, dams are not usually built to accommodate the most extreme floods, which are also the most likely to pose a threat to life and property. Furthermore, minor flooding presents benefits—for example, plant nutrient delivery to soils in the floodplain—that are eliminated by dams. Regarding navigation, since dams are expensive and silt up rather quickly, requiring dredging, in certain cases dams are not the most cost-effective means of improving navigation. In other words, dams are expensive to build, and because they slow down a river's flow, they cause sedimentation upriver of the dam as solids are deposited from suspension in the water column, resulting in increased need to dredge away this sediment to maintain a waterway navigable to large ships. Furthermore, due to the fees and time lost passing through locks, it can be uneconomical for smaller businesses and/or local commerce, especially considering that river transportation is dominated by a few large corporations. For a more detailed discussion, see Patrick McCully's *Silenced Rivers: The Ecology and Politics of Large Dams* (New York: Zed Books, 2001).

12. For a fascinating critical history, see Marc Reisner's *Cadillac Desert: The American West and Its Disappearing Water* (New York: Penguin Books, 1993).

13. Intriguingly, increased seismic activity and even changes in the Earth's rotation are also associated with dams and their reservoirs. NASA geophysicist Dr. Benjamin Fong Chao found evidence that large dams cause changes in the earth's rotation due to the shift of gravity entailed by moving water from oceans to reservoirs. Because of the number of dams that have been built, the Earth's daily rotation has apparently changed by eight-millionths of a second since the 1950s. Chao said it is the first time human activity has been shown to have a measurable effect on the Earth's motion. For more information, see Benjamin F. Chao, "Anthropogenic Impact on Global Geodynamics Due to Water Impoundment in Major Reservoirs," *Geophysical Research Letters* 22 (1995): 3533–3536.

14. Federal hydropower projects are a separate matter in terms of federal oversight and are hence beyond the scope of present inquiry. These are typically larger scale, such as the large-scale projects in the western United States, including those constructed by the Army Corps of Engineers, and those across the Tennessee Valley, regulated by the Tennessee Valley Authority, including the Little Tennessee River's Tellico Dam, famously delayed in 1973 by the snail darter controversy, which saw the 1973 Endangered Species Act put to an early test. The Gorge Dam is private and therefore falls under the scope of FERC's authority.

15. However, there have been only three members under the Bush and Trump

administrations, two of whom were Republican, and considering that FERC decides by a simple majority vote and the current composition of the committee, FERC's impartiality condition is arguably not currently being satisfied.

16. John D. Echeverria, Pope Barrow, and Richard Roos-Collins, *Rivers at Risk: A Concerned Citizen's Guide to Hydropower* (Washington, DC: Island Press, 1989), 21–22.

17. "Encyclopedia of Cleveland History: Cuyahoga River," Case Western Reserve University, https://case.edu/ech/articles/c/cuyahoga-river.

18. "Ohio and Erie Canal National Heritage Corridor: Industry," National Park Service, https://www.nps.gov/nr/travel/ohioeriecanal/industry.htm.

19. Tony Long, "June 22, 1969: Umm, the Cuyahoga River's on Fire . . . Again," *Wired Magazine* (blog), June 22, 2011, https://www.wired.com/2011/06/0622cleveland-cuyahoga-river-fire-burns-again/.

20. Jennifer Latson, "The Burning River That Sparked a Revolution," *Time* magazine, June 22, 2015, https://time.com/3921976/cuyahoga-fire/.

21. "History of the Clean Water Act (CWA)," Environmental Protection Agency (EPA), https://cfpub.epa.gov/watertrain/moduleFrame.cfm?parent_object_id=2571.

22. "Cuyahoga River Remedial Action Plan," EPA, updated July 18, 2018, https://www.epa.gov/great-lakes-aocs/cuyahoga-river-remedial-action-plan.

23. "Cuyahoga River AOC," EPA, updated July 20, 2020, http://www.epa.gov/great-lakes-aocs/cuyahoga-river-aoc.

24. "Dam Restoration Project," The City of Kent, Ohio, https://www.kentohio.org/409/Dam-Restoration-Project.

25. Mark Urkcki, "The Unexpected Consequences of Cuyahoga River Dams," *IdeaStream*, April 26, 2019, https://www.ideastream.org/news/the-unexpected-consequences-of-cuyahoga-river-dams.

26. Urkcki, "Unexpected Consequences."

27. "Gorge Dam: Cuyahoga River Area of Concern."

28. However, such figures about energy generation capacity can be highly misleading because a river's flow varies substantially throughout the year. The available flow is *much* lower in the summer months, which is when electricity demand is at its highest due to air conditioners, which strongly suggests that peak operational capacity is a mismatch with peak local demand.

29. A. Ritchie and J. R. Steiger, *Soil Survey of Summit County, Ohio* (Washington, DC: Soil Conservation Service, U.S. Department of Agriculture, 1974).

30. "Formal Objection to Metro Hydroelectric Company, Project Number P-12484," Metro Parks, Serving Summit County, https://www.ferc.gov. Metro Parks had the following

to say: "Many of the trees on this side of the Gorge valley . . . indicate old growth conditions. Regeneration of this already scarce habitat would take more than 100 years and could not possibly be mitigated. . . . Soil disturbance and removal of old growth tree canopy would also create conditions that are highly favorable for the introduction and domination of invasive plant species . . . [which are] the single greatest threat to the biodiversity within the Metro Parks. . . . Despite this obvious threat associated with the proposed project, there is no mention of this very real impact anywhere in the PAD."

31. Ohio Environmental Protection Agency, *Biological and Water Quality Study of the Cuyahoga River and Selected Tributaries*, vol. 1, 1996, https://www.epa.ohio.gov.

32. The following organizations objected or voiced concern over the project: Metro Parks Serving Summit County, Ohio Environmental Protection Agency, Ohio Environmental Council, Friends of the Crooked River, Northeast Ohio Watershed Council, American Rivers, Rivers Unlimited, Keel Haulers Canoe Club, American Whitewater, City of Cuyahoga Falls, Cuyahoga Valley Communities Council, Summit Soil and Water Conservation District, Summit County Council, Cuyahoga Remedial Action Plan, Cuyahoga River Navigator, Cuyahoga Valley National Park, Geauga County Metroparks, Friends of the Metro Parks, Akron Garden Club, Kent Environmental Council, and Friends of Wetlands.

33. Ohio Environmental Protection Agency, *Total Maximum Daily Loads for the Lower Cuyahoga River*, September 2003, 17, https://tinkerscreek.org/wp-content/uploads/2016/05/TMDL.pdf.

34. Tetra Tech, *Feasibility Study for the Removal of Gorge Dam*, September 21, 2015, https://dokumen.tips/reader/f/feasibility-study-for-the-removal-of-the-gorge-dam-reportpdf-feasibility-study.

35. Urkcki, "Unexpected Consequences."

36. Bob Downing, "No One Knows Who Owns the Gorge Dam on the Cuyahoga River between Akron and Cuyahoga Falls," *Akron Beacon Journal*, October 12, 2015.

37. Jennifer Conn, "Plan Unveiled to Bring Down the Gorge Dam by 2023," WKSU, April 10, 2019, https://www.wksu.org/post/plan-unveiled-bring-down-gorge-dam-2023.

38. Larry Hickman, "John Dewey's Pragmatic Naturalism: Nature as Culture," in Light and Katz, *Environmental Pragmatism*, 69 n. 2. Hickman notes that early critics included Bertrand Russell in the UK, Theodor W. Adorno and Max Horkheimer in Germany, and even his contemporary George Santayana, who presented Dewey as "the devoted spokesman of the spirit of enterprise, of experiment, of modern industry" whose philosophy was "calculated to justify all the assumptions of the American society."

39. James Gouinlock, *The Moral Writings of John Dewey: A Selection* (New York: Hafner Press,

1976), xix.

40. Paul Thompson, "Pragmatism and Policy: The Case of Water," in Light and Katz, *Environmental Pragmatism*, 187.

41. See John O'Neill, "The Varieties of Intrinsic Value," *The Monist* 75 (1992): 119–137. This is a helpful analysis of this topic.

42. John Dewey, *Quest for Certainty* (New York: G. P. Putnam's Sons, 1929).

43. Some environmental ethicists who are not explicitly pragmatists also hold this view. Dale Jamieson seems to endorse this position. See his "Animal Liberation Is an Environmental Ethic," *Environmental Values* 7, no. 1 (February 1998): 41–57, at 23: "I speak of 'intrinsically valuing' rather than 'intrinsic value' because it makes clear that the intended distinction is in the structure of valuing rather than in the sorts of things that are valued. We intrinsically value something when we value it for its own sake. Making the distinction in this way also makes clear that one and the same entity can be valued both intrinsically and non-intrinsically at different times, in different contexts, by different valuers, or even by the same valuer at the same time. For example, I can intrinsically value Sean (i.e., value her for her own sake) yet non-intrinsically value her as an efficient mail-delivering device (i.e., for how she conduces to my ends). Collecting these distinctions we can entertain the possibility that the content of our values may include our intrinsically valuing an entity that is of derivative value, and that this valuing may be urgent and intense, even trumping something of primary value. The obvious candidates for satisfying this description are works of art. Many of us would say that the greatest works of art are very valuable indeed. We value them intrinsically, yet ultimately an account of their value devolves into understandings about their relations to people (e.g., artists, audiences, potential audiences, those who know of their existence, etc.)."

44. There is a worrying possibility of anti-intellectualism here, but I take it that the pragmatist point is that ivory tower debate *by itself* is not particularly good at advancing the preservation of the value of the very object in question. Resolving the tension between academic inquiry and environmental activism is beyond the scope of this inquiry, but it is a serious question for environmental philosophy.

45. Matthew Festenstein, "Dewey's Political Philosophy," *The Stanford Encyclopedia of Philosophy*, Winter 2019, https://plato.stanford.edu/entries/dewey-political/.

46. Gouinlock, *Moral Writings of John Dewey*, xx.

47. For a brief overview, see Alvin Goldman and Cailin O'Connor, "Social Epistemology," *The Stanford Encyclopedia of Philosophy*, Fall 2019, https://plato.stanford.edu/archives/fall2019/entries/epistemology-social/.

48. Udall v. Federal Power Commission, 387 U.S. 428 (1967), at 450.

49. Electric Consumers Protection Act of 1986, Section 4(e), emphasis added.

50. Federal Power Act of 1935, Section 10(a)(1), emphasis added.

51. See "Meeting Minutes of July 27, 2005, Meeting on the Metro Hydroelectric Project," Federal Energy Regulatory Commission, available at https://www.ferc.gov.

52. "Mission Statement," Cuyahoga River Area of Concern (AOC), https://www.cuyahogaaoc.org.

53. "Study Plan Determination for the Metro Hydroelectric Project," Federal Energy Regulatory Commission, Docket # P-12484-001, https://www.ferc.gov.

54. "February 22, 2006, Letter to Federal Energy Regulatory Commission," Ohio Environmental Protection Agency, https://www.ferc.gov.

55. Ohio Environmental Protection Agency, "22 February 2006 Letter."

56. "Gorge Dam Hydroelectric Project," Friends of the Crooked River, 2020, http://cuyahogariver.net/gorge-dam-hydroelectric-project/.

57. John D. Echeverria, Pope Barrow, and Richard Roos-Collins, *Rivers at Risk: A Concerned Citizen's Guide to Hydropower* (Washington, DC: Island Press, 1989), 21–22.

58. "Study Plan Determination for the Metro Hydroelectric Project."

59. For example, see Sabine Maasen and Peter Weingart, eds., *Democratization or Expertise? Exploring Novel Forms of Scientific Advice in Political Decision Making* (Dordrecht, Netherlands: Springer Press, 2006).

60. John Dewey, *The Public and Its Problems* (Athens, OH: Swallow Press, 1927), 208.

61. Dewey, *The Public and Its Problems*, 208–209, 207.

62. David H. Guston, *Between Politics and Science: Assuring the Productivity and Integrity of Research* (New York: Cambridge University Press, 1999).

63. Heather Douglas, "Inserting the Public into Science," in *Democratization or Expertise? Exploring Novel Forms of Scientific Advice in Political Decision Making*, ed. Sabine Maasen and Peter Weingart (Dordrecht, Netherlands: Springer, 2005), 157.

64. Amy Kober, "American Rivers Names Cleveland's Cuyahoga 'River of the Year,'" American Rivers, https://www.americanrivers.org/conservation-resource/american-rivers-names-clevelands-cuyahoga-river-of-the-year/.

SUGGESTED READINGS

Cully, Patrick M. *Silenced Rivers: The Ecology and Politics of Large Dams.* New York: Zed Books, 2001.

Dewey, John. *The Public and Its Problems*. Athens, OH: Swallow Press, 1927.

Grossman, Elizabeth. *Watershed: The Undamming of America*. New York: Counterpoint, 2002.

Light, Andrew, and Eric Katz, eds. *Environmental Pragmatism*. New York: Routledge, 1996.

McDonald, Hugh P. *John Dewey and Environmental Philosophy*. Albany: State University of New York Press, 2004.

What Ohio Can Teach Us about Effective and Ethical Ecological Restoration

Justin Donhauser

O hio is unique because of its diversity of historical bioregions, including wetland, grassland, mountain, and various kinds of forest and aquatic bioregions. At the same time, like much of the broader Midwest, the state contains large areas that were once converted to farmland and land for commercial and residential use. A large percentage of that land is fertile land that is no longer used at all. These factors make many parts of Ohio near perfect candidates for many kinds of valuable ecological restoration projects.

This chapter examines three restoration strategies employed across Ohio. These include instances of defunct farmland restoration, assisted colonization of predators (for example, black bears and coyotes), and "novel ecosystem" engineering in and around wetlands and disconnected watersheds. For each of the examined strategies, specific case studies in the state are discussed to highlight positive and negative tradeoffs of each, as well as to gather generalizable insights for how to effectively and ethically employ similar restoration strategies elsewhere. The discussions of those case studies and more general strategies are prefaced by a brief presentation of the virtue-ethical framework through the lens of which I will evaluate the selected cases. The anthropocentric values that motivate restoration efforts are often at odds with non-anthropocentric values that many have pointed out must be preserved to effectively address issues associated with restoration efforts, such as global climate change and environmental justice.[1] That tension is one between trying to maximize goods for humans, such as maximum crop yield or "ecosystem services" to human populations, and maximizing forms of natural composition and functionality that may better serve nonhuman populations, communities, or systems. This chapter goes some way toward showing that these seemingly competing sorts of aims are not necessarily at odds. It also provides some generalizable insights, and a normative view for framing responses to environmental problems that reconciles the relevant anthropocentric and non-anthropocentric values. For each examined case study, I engage with that virtue-ethical perspective to reflect on how similar efforts could aid in responding to global climate change and biodiversity loss, as well as reflect on how such strategies might help address environmental and ecological justice concerns elsewhere.

A Virtue-Ethical Environmental Ethics?

To contextualize the subsequent discussion, I'll outline here a virtue-ethical way of framing environmental protection and management that I have defended in another work.[2] The virtue-centered view that I endorse suggests striving for interventions that can enhance well-being in a place by making one more a part of it. In this endorsed view, "part of what it is to live a good life is to contribute in the right way to ongoing social-ecological functioning of the systems to which we belong."[3] So, interventions that permit one to live well in their environment and satisfy holistic well-being (for example, sustainable agricultural practices) would be more virtuous than those that satisfied expressed preferences (for example, maximum-yield agricultural practices with adverse environmental side effects). In other words, we would be better people if we pursued those interventions that permitted us to live well in our environment(s), rather than those that simply fulfilled our preferences. Accordingly, the endorsed view does not imply that humans ought to pursue environmental engineering and management strategies as a means of controlling nature and natural processes. This is again because overall, holistic well-being requires functional integration in situ (that is, in each particular place and situation). As the environmental ethicist Holmes Rolston

III has put it, "Our virtues defend the organic self, but they also stretch it out to integrate it into its place."[4] What this means is that whatever environment an individual person or organism or population finds itself living in, that individual's well-being can be enhanced or diminished by the extent to which it utilizes and contributes to its environment and the various sorts of resources that a particular environment can provide.

In accordance with this line of thinking, the virtue ethicist's conception of "external goods" is particularly helpful for evaluating environmental protection and restoration initiatives and technologies. External goods—for example, wealth, friendship, and peace in society—are simply goods outside of an individual that can help one to actualize their potential for forms of well-being and flourishing in a particular place; they are, in other words, functional background conditions that can help enable one's ability to actualize his or her well-being capacities in a place.

External goods can be better or worse in kind and quality. So, for instance, although having access to drinking water and air are minimally sufficient as external goods in kind, it is clear that there could be better or worse kinds of water or air access. Accordingly, better external goods have qualities, and a relationship with one engaging in such external goods, that promote and support his or her well-being in a holistic sense.[5] In this view, good environmental interventions can be those that help one live a healthier life by biological standards achieved through protecting or enhancing environmental resources. Yet, in this view, good interventions cannot be just any interventions that fulfill our preferences.

Notably, which considerations factor into how any individual could live a functionally better or worse life will differ in different locations, at different times, given the different environmental conditions and potential challenges that exist in each particular place. A virtue-centered view is highly adaptive and lends itself to thinking about how well-being may be realized differently at different times and places. This is because it recognizes that individual well-being is dependent on collective well-being, collective good functioning of many sorts; our well-being and flourishing is sensitive to our existence as parts of "ecosystems in which interdependency, interpenetration, and causal entanglement between organisms and their environments are the norm."[6]

The view also suggests that we have responsibilities, or at least that it is virtuous, to develop and carry out environmental interventions that would protect capacities for well-being and flourishing or enable new ones.[7] In other words, "new forms of human goodness entail new virtues, just as genuinely new virtues implicate new forms of the good life."[8] Hence, if there are environmental interventions that would bring about goods in the holistic sense described so far, we have virtues directed at pursuing those interventions.

Extending this thinking just a bit further, it is also instructive, and useful when evaluating environmental interventions, to recognize virtues directed toward nonhuman species and nature and their external goods and well-being interests. Ian Smith builds on John O'Neill's conception of "biological good" to formulate an objective conception of capacities for well-being and flourishing in nonhuman nature. In Smith's formulation, external or extrinsic (which he calls "instrumental") biological goods for any biological individual include environmental conditions that enable it to realize its natural capacities and flourish, and internal (which he calls "intrinsic") goods of any individual are that individual's natural capacities. Hence, for any kind of biological entity (organism, population, community, or species) to flourish, or be better or worse, is for that individual to realize its biological capacities and potential or not to do so.[9]

This conception of things may seem odd at first, but it is easy to see that things can go better or worse for forms of nonhuman nature. For example, a biological population tending toward local extinction, via the members of that population dying and failing to reproduce, is of course bad for that population. Conversely, its persisting is good for that particular population. And though Smith and others who share in this thinking may not extend this thinking, this simple line of reasoning allows us to extend moral considerability to many kinds of nonhuman nature.[10] The reason for this is that protecting and enhancing the well-being capacities of nonhuman nature can be partially constitutive of human well-being and flourishing. Doing so is virtuous, and good in virtue-ethical terms, when it is good for us in at least four ways.

First, protecting and promoting the well-being of many forms of nonhuman nature can be partially constitutive of our biological well-being.[11] So, since many forms of nonhuman nature are good for our survival and well-being, it is virtuous to protect and promote their well-being and capacities to fill their natural roles as functional parts of our environments. Second, since humility is a virtue, we ought to pursue initiatives to restore and enhance the well-being capacities of many forms of nonhuman nature that anthropogenic actions have degraded or destroyed.[12] Third, using our uniquely evolved reasoning capacities to conclude that forms of nonhuman nature have their own goods owing to their capacities for well-being, and taking actions to promote their well-being can contribute to our psychological well-being.[13] And finally, insofar as pursuing justice is a virtue, it is arguable that protecting or restoring the forms of nonhuman nature is virtuous in cases where that entails rectifying environmental or "ecological" injustices. As Schlosberg says, speaking of ecological injustices, "it is a lack of flourishing that is indicative of injustice," and in order to dissolve ecological injustices, what must be remedied is "the absence of specific capabilities that produce flourishing."[14] Accordingly, I follow Schlosberg further in concluding that

> Flourishing is not an element that relates only to humanity, nor is it an element based in human life that we simply apply to nature along the lines of similarity. [Well-being capacities] include what is necessary for functioning and flourishing of human and nonhuman alike; it is an integral aspect of the living process.[15]

Reasons to Favor a Virtue-Ethical Perspective

To this point, I have outlined the main threads of the virtue-ethical view I'll use to evaluate different environmental restoration strategies. I'll now provide motivation for that latter project by explaining some reasons the endorsed virtue-ethical view is attractive. Certain general upshots of the outlined virtue-ethical view can be seen by considering how it avoids problems faced by alternative views.

To begin, there are utilitarian views that see ethically good restoration and preservation strategies as those that maximize useful goods—generally for humans, but in some formulations also for nonhuman organisms. And then there are deontological views that basically see ethical strategies as those that would always be morally right for humans, and in some formulations also nonhuman organisms, in all contexts. Notably, these alternatives to a virtue-ethical perspective do not provide clear guidance in particular situations unless they are grounded in some guiding set of values that are arguably impossible to agree upon and are likely to be incompatible with how values might change across time or across different contexts.[16] In fact, in the cases discussed below we will see that environmental problems, and the need for restoration and protection efforts, are often due to changing values or conflicts between different sorts of values. Because utilitarian and deontological views basically provide systems for determining whether any action is right or wrong, they can also have difficulty being adaptive and responsive to the context-sensitivity of many questions regarding restoration and natural resource management. This is because a specific restoration and protection initiative may not be clearly right or wrong but most often simply one feasible practical response among many other practically viable responses.

Asking whether it is *right* or *wrong* to pursue any environmental restoration or protection project is asking a question to which there can be no certain answer. This is first because, to the extent that such questions of ethicalness can even be answered, such answers will always depend on the context at any particular time. This is to say that what could be right at one time could be wrong at another. For example, below I will discuss predator reintroductions. Whether such reintroductions are right or wrong depends entirely on the context and environmental conditions at the time of the reintroduction. What's more, it seems as though rightness and wrongness are "the wrong sorts of metrics" for evaluating restoration and protection initiatives, since such initiatives are designed to bring about environmental goods and protect environmental values, even though they may have negative or uncertain impacts in certain contexts.[17] So, again utilitarian and deontological theories provide limited guidance for evaluating restoration and protection strategies.

As I have argued elsewhere, "Virtue-centered views do not face such problems because they are adaptive. Whatever the context and the question, any consideration is guided by evaluating the extent to which the action in question will enable and promote well-being and flourishing."[18] This is not to deny that some variations of utilitarian and deontological ethical systems may have resources for helping address questions about environmental restoration efforts. It is simply to say that a virtue-centered view is naturally better aligned with the context sensitivity of such efforts. Furthermore, utilitarian and deontological alternatives arguably have problems providing clear guidance in practice. As Vallor writes:

> [We must make] sense of each concrete situation encountered, and give an appropriate response. A successful moral response is distinguishable from a failed or inappropriate response *in practice*, and the reasons behind the success of that response can always be articulated after the fact. But the difference between moral success and moral failure can rarely, if ever, be deduced in advance from *a priori* principles.[19]

A virtue-centered view also avoids three well-known difficulties with these competing views. First, a virtue-ethical view can avoid problems with a commonplace anthropocentric utilitarianism that sees valuable ecological functions and worthy environmental problem responses "as those that serve current human preferences as expressed by market behavior."[20] Because it concentrates on promoting well-being and flourishing, and conceives of human and nonhuman well-being holistically, a virtue-ethical view does not assume that goods can only be those that can be treated as quantifiable instrumental goods. Since a central difficulty for utilitarian views is that some things that support human well-being altogether resist quantification (for example, cultural, spiritual, and psychological goods), this makes a virtue-centered view easier to use in practice. Notably, a virtue-centered view also puts constraints on following expressed preferences and market behaviors, since that can lead us down unvirtuous paths. For instance, expressed market preferences in agriculture and livestock production could indicate a high demand for killing nuisance predators like coyotes that can harm cash crops and livestock. Yet, arguably, hunting coyotes to extinction and thus harming broader ecological functionality is unvirtuous behavior.

Second, the endorsed view does not warrant the human domination and control of natural processes that is warranted by certain utilitarian and deontological views. And arguably it is unvirtuous, and psychologically bad for our human well-being, to engage with an ideology of such domination and control. As the environmental ethicist Holmes Rolston III has written:

> Each species, each individual, sets a boundary between itself and the rest of nature, and in humans that discontinuity is enormously greater than elsewhere. This spirited agency is the distinctly human genius, the human virtue, wrested from nature. . . . We maintain our being by being over against nature. . . . [Yet, we at once] flourish with our landscapes, with trees and grass, flowers and gardens, lakes and sky. . . . Most of us identify so with some countryside that we get a lump in the throat when we must leave it or when we return after an absence. We have deep affections toward persons and communities, but our affections toward the city, per se, are often exceeded by those that we have toward the landscape. . . . Our virtues defend the organic self, but they also stretch it out to integrate it into its place.[21]

Accordingly, and third, the virtue-ethical view sees that goods for humans and nonhumans are deeply context-sensitive. The view therefore avoids problems that stem from the universalist core tenets of deontological views, as deontological views suggest that good solutions or actions are those that would be right in all contexts. The virtue-ethical view also outstrips deontological views because they emphasize *individual* autonomy and rights. This is because, according to the virtue-ethical view, individual well-being is dependent on collective well-being of many sorts. In other words, human well-being and flourishing is sensitive to our being parts of "ecosystems in which interdependency, interpenetration, and causal entanglement between organisms and their environments are the norm."[22] Reflection on this point leads to the conclusion that we have ecologically directed virtues, and to thinking about how different environmental protection and restoration strategies could best promote well-being and flourishing.

Farmland Rewilding

Against this background, I now turn to examine specific case studies in Ohio to highlight positive and negative tradeoffs of each and gather generalizable insights for how to effectively and ethically employ similar restoration strategies. To begin, let's consider farmland rewilding initiatives in the state. Of note, farmland rewilding is about more than just restoring land to its "natural" state and promoting biodiversity in an area. It can also serve to offset the environmental damages and risks that can be caused by certain kinds of farming practices.

For example, the Midwest in general and certain parts of Ohio in particular are home to large-scale concentrated animal feeding operations (CAFOs)—often conventionally called "factory farms." CAFOs include, but are not limited to, dairy, beef, swine, and poultry facilities, which all have different footprints and environmental impacts; but, given conventional practices, they all have negative environmental impacts and expose adjacent populations to health risks. Each year, CAFOs produce hundreds of millions of tons of manure in the United States that cause corrosive, and sometimes toxic, levels of CO_2 and methane emissions. This can cause water and air contamination that impact the health of surrounding residents.[23] For instance, numerous studies examine concerns about CAFO siting and increased illness in children and elderly in adjacent communities due to toxic emissions and airborne illnesses caused by CAFOs.[24] As one of the country's largest egg producers, Ohio has continued to grow its numbers of CAFOs.

CAFOs are sited in places where there are sufficiently large amounts of land for livestock farming, production, and access to markets. So, they are generally sited on rural land adjacent to cities and means of transport. Accordingly, they've historically been disproportionately sited near areas with large minority-group populations and low-income families. A study of CAFOs along Ohio's western border by Lenhardt and Ogneva-Himmelberger finds that "black and Hispanic populations, as well as households with relatively low incomes, are disproportionately exposed to CAFOs compared to other populations," and that persons in those populations were statistically likely to "experience reduced quality of life as a result of CAFO proximity."[25] What's more, CAFOs can severely impact the regional economy by putting smaller farms (for example, "family farms") and related businesses out of business.

Of course, there are smaller-scale sustainable livestock as well as other farming practices that avoid the negative impacts and risks of CAFOs and would even help offset them. These alternative practices could serve the needs of localized communities, thereby attending to some of those communities' well-being interests. Additionally, they could be carried out in such a way that they would also enhance and protect the well-being interests of livestock populations as well as those of adjacent natural populations.[26]

Case

To push forward our discussion, let us now begin to consider farmland restoration and rewilding practices. An exemplary project, piloted in 2004, sought to bring back wet prairie, swamp forest, and oak savanna areas to 16.2 hectares of prior farmland in the northwestern edge of the historical "Great Black Swamp" region. The effort's primary aims were to restore biodiversity, especially rare and dwindling plant species, and to provide buffer zones that would reduce the nutrient load going into Lake Erie, which is beneficial because it prevents the emergence of algal blooms and other microplant overgrowth that can lead to eutrophication and the destruction of entire ecological communities in lake systems. In sum, the outcome of the effort was the reestablishment of eighty species in the prairie areas, including six endangered species, 435 trees in the forest areas, and a large reduction (even zero from April to December) in nutrient outflow from the wetland area.[27] What's more interesting than these successes for the purposes of this chapter are the lessons to be gleaned from what happened throughout the project.

To begin, the successes of this project clearly demonstrate that wetlands and historical wet prairie, wetland forest, and swamp regions can provide crucially important ecological functions and services to human populations. Showing this, of course, helps ameliorate the fact that "wetlands of any type are still often viewed negatively in the former Great Black Swamp," and can thus reshape community values and foster stewardship of the historical ecology of the region.[28] What's more, the particular ecological functions and functional relations to human and nonhuman populations hold some generalizable lessons when looked at through the virtue-ethical view outlined above. I'll now present some of the key functions highlighted in this project and then discuss some relevant generalizable lessons.

Foremost, the restored prairie, wetland forest, and swamp regions functioned together to control the hydrology of the broader region.[29] This allowed for a conceptualization of the entire project region as a functional network or system. Roughly ten years after the inception of the project, after the functional regions had time to "take root," so to speak, over 99 percent of the precipitation water that fell on the site was retained within the site's subsurface base. This retention of water served to make the site a hydrologic buffer between the remaining farmland on one side and the Maumee River on the other. Monitoring at the nearby headwaters of a small tributary to the Maumee River showed *no* drainage and thus no runoff from upstream farmland. So, the site was situated in a "very favorable location for water and nutrient retention," as its "high wetland to watershed area ratio makes it efficient at removing nutrient outflow that would otherwise contribute to Lake Erie algae bloom problems."[30] This is not at all inconsequential, as toxic algal blooms have been a huge problem in Lake Erie, even causing Toledo, Ohio, to shut down the municipal water supply in 2014.[31]

At the same time, researchers found that the site provided some fairly standard challenges, in that it could not continue to function autonomously without human intervention and management. Although they succeeded in establishing native plant cover, control of invasive species was an ongoing challenge that required continual intervention in the prairie region of the project. They also projected that long-term sustainability of the prairie and savanna areas would require periodic controlled burns. Additionally, they reported a high seedling mortality rate in some other regions due to the soil composition (clay) and grazing from deer and rabbits that meant that management and reseeding would need to extend decades, "beyond the typical career length," to establish the historical forest canopy.[32] Despite these hurdles, farmland restoration can have broad-reaching positive systemic impacts. Further, there is research on efforts such as that highlighted above suggesting that farmland restoration can often serve as a sustainable solution to many problems in an entire watershed.[33]

Discussion

In line with the virtue-ethical position outlined in the previous section, the functional benefits and even practical hurdles can all be seen as positive features of projects like the one discussed above. Wetland restoration projects like the one above are naturally geared toward protecting and promoting the well-being of many forms of nonhuman nature. This in turn can be partially constitutive of our (human) well-being. Perhaps most obviously they can improve water, air, and soil quality in adjacent regions by serving as a control on hydrology, since they slow and filter water flowing through to waterways further down the watershed. They can also provide habitat and new range for many nonhuman species. So, such efforts are arguably virtuous in that they aim to protect and promote the well-being of humans and nonhuman species by providing mechanisms for each to fill roles as functional parts of their environment.[34]

The aims of restoration projects generally also serve to exercise forms of humility, since such initiatives, in effect, try to restore and enhance the well-being capacities of many forms of nonhuman nature that human land uses and modifications have degraded or destroyed. Accordingly, though conservationists don't put it this way, restoration initiatives see forms of nonhuman nature as having their own capacities for certain kinds of well-being and consist in taking actions that promote such well-being. Arguably, because of this, engaging in such initiatives has psychological benefits for those involved in them.[35]

Efforts like the above-mentioned wetland restoration can also be seen as virtuous initiatives in that they can entail rectifying environmental and/or ecological injustices. Intentional, or specifically directed, "environmental justice" has tended to focus on rectifying environmental injustices experienced by human populations.[36] On the other hand, ecological justice initiatives pursue justice for forms of nonhuman nature—for example, enhancing the well-being capacities of some animal population that has been diminished due to human-driven environmental changes.[37] Restoration efforts can serve as a means of addressing environmental-justice concerns effectively while simultaneously addressing concerns of ecological justice. For example, by controlling—and in the above case stopping—the hydrologic throughput of runoff contaminants from farms, such wetland restorations can help mitigate the environmental justice issues concerning CAFOs while at once providing habitat and ecological goods to enhance the well-being capacities of numerous forms of human and nonhuman nature. In fact, this would mitigate environmental racism issues stemming from problems like CAFOs being disproportionately sited near areas with large minority populations.

A final score on which restoration efforts can be evaluated as more or less virtuous concerns how well they enhance the well-being capacities of relevant stakeholders (that is, people invested in and/or functionally impacted by a restoration effort) and how well they facilitate their functional integration in a place. Recall that the endorsed virtue-centered view recommends striving for interventions that evaluate goods holistically and thus sees part of what it takes to enhance human well-being in a place as making one a more functional part of that place.[38] By helping reshape community values and fostering stewardship of the historical ecology of a region, as in the above-discussed wetland restoration, restoration initiatives can serve as a means of functionally integrating people in this way. Accordingly, the types of sustained and cyclical patterns of intentional maintenance that are often required in restoration projects can serve to reinforce and enhance such functional integration along with increasing intentional appreciation in a place.

Assisted colonization and novel ecosystem engineering projects like those I will now discuss can also be evaluated on all of these scores through the lens of the endorsed virtue-ethical perspective. So, I will not detail all of the ways that those efforts can be seen as more or less valuable, or virtuous, in line with the view. I have instead selected cases that highlight other aspects of conservation and restoration efforts that provide generalizable lessons in order to extend the considerations of this chapter further, and provide some further insight into how the virtue-ethical perspective is useful for evaluating restoration and engineering efforts of different sorts.

Assisted Colonization of Small Predators

Cases of "assisted colonization" of small predators in areas of Ohio provide some interesting generalizable insights regarding certain social and public-relations-type aspects of effectively realizing conservation strategies. Although there is debate about the best terminology for discussion of this restoration strategy, it is called either "assisted colonization," "assisted migration," or "managed relocation."[39] I have chosen to call it Assisted Colonization (AC) in this chapter. ACs are efforts to move populations of a given species from regions where they are threatened with local extinction—for example, by habitat loss or predation by invasive species—into regions typically outside their historical range. Such efforts aim to locate—or in some cases to relocate—populations where they can be more functionally resilient in an effort to prevent their functional extinction.

As a brief point of clarification about terminology, in this section I will discuss ACs that many would see as conventional population "reintroductions." Notably, "reintroductions" do not always involve relocating wild populations. For example, they often use animal populations that are bred in captivity, and often involve putting such captive-breed populations into habitats that are historically native habitats for populations of the salient species. Of course, doing this is different than moving a population from a habitat where it isn't historically native and helping that population colonize a (potentially new) more historically viable type of habitat. Still, in this chapter, I will call both "reintroductions" and what some may distinguish as assisted migration, colonization, or relocation efforts all ACs for simplicity—and because, in my view, "reintroductions" (as just described) are one kind of assisted colonization.[40]

There are two primary reasons for the need for ACs. First, human land use has reduced many populations such that many species exist in small, fragmented populations, and are therefore vulnerable to environmental threats and lack long-term resilience and survivability.[41] The other increasingly threatening environmental situation results from impacts on a given environment due to anthropogenic climate change. There are various arguments to compel saving species from extinction. Such arguments generally appeal to different values that species possess, including kinds of instrumental value (such as serving as food or playing important roles in ecosystems) and kinds of intrinsic value (such as having value independent of their usefulness to humans because they have certain kinds of goals and well-being interests).[42] Doing ACs for any of these reasons is consistent with exercising humility and the promotion of human and nonhuman well-being interests. These reasons represent the motivations for pursuing such initiatives according to the virtue-ethical view I've endorsed. Discussing two AC projects in Ohio will help illuminate the importance of certain public education aspects of conservation efforts that can make them more effective, in addition to highlighting the point that anthropocentric conservationist and restorationist efforts are consistent with non-anthropocentric efforts.

Cases

Serfass, Bohrman, Stevens, and Bruskotter chronicle the many negative and positive portrayals of the North American river otter (*Lontra canadensis*), and the impacts that public perception can have on the realization and long-term success of AC initiatives to protect the otter.[43] Otters are native to almost every U.S. state, but have almost disappeared and have become extinct in much of their historical range due to trapping in the early twentieth century, anthropogenic disturbances to their riparian habitats, and pollution.[44] As Serfass et al. state, "media portrayals about otter reintroductions in the United States have ranged from descriptions of otters' playful nature and positive role in aquatic ecosystems, to negative depictions of their feeding habits."[45] Notably, these portrayals have significantly impacted AC efforts.

For instance, otters were reintroduced in Ohio between 1986 and 1993. Yet, shortly after those efforts many residents complained about the otters' impacts on fish populations. The AC efforts were even called "almost too successful" in some media, and other sources even made claims that "otters are overrunning Ohio."[46] By 2005 a trapping season for otters was instated and has been expanded.[47] This of course undermines AC efforts to bring the otter back in Ohio, since the otter was not "overrunning Ohio" from a historical or ecological perspective. Contrast cases help further illustrate how crucially important public perception is. In Pennsylvania, for example, there was little public attention or criticism after

AC efforts to reintroduce otters, even though it was one of the very first states to pursue such reintroductions. The Pennsylvania River Otter Reintroduction Project began in the early 1980s and has been very successful in comparison to the efforts in Ohio.[48]

Other ACs of small predators also show parallels to the otter case. For instance, consider muskrat and other woodrat reintroductions in Ohio. Serfass considers what can be learned from reintroductions including "the release of 44 woodrats from Ohio and Kentucky at a site in Ohio during 1983 and 1984."[49] He recommends numerous preliminary assessments that can be used to establish justifications and protocols to help ensure the success of ACs.

Serfass recommends focusing primarily on assessments to determine how to realize effective public outreach and education, with a focus on building public support for AC projects prior to implementation planning. He argues that outreach "represents the most viable approach for achieving positive, long-term outcomes for important conservation issues, including the re-establishment of species by reintroduction."[50] Serfass notes, drawing from vast experience, that conservation professionals most often neglect, or are just completely oblivious to the very positive long-term impacts outreach initiatives can have. Indeed, the value of public outreach and education has been demonstrated in numerous AC projects besides the small predator introductions mentioned above.

Notably, successes have been achieved through efforts to make a target species for reintroduction via ACs appealing to the public. Successes have been seen in efforts to introduce larger predators that the public find impressive, such as wolves, for instance.[51] Yet, even in the case of river otters, successes have been seen when the public comes to see them as cute or "charismatic."[52] Serfass suggests garnering public support by emphasizing any unique characteristics or behaviors of a target species for ACs (such as the muskrats' building techniques and use of middens to store wasted shells from eating oysters), and organizing simple educational events that showcase the target species' natural history and conservation status.[53]

Discussion

In accordance with the endorsed virtue-ethical view, we can of course make judgments about what happened in the small predator introduction cases in Ohio. For instance, one could easily argue that the otter reintroduction would have been better, more virtuous, had care been taken to work against the uninformed negative public perception of otters in that case. However, what I find more interesting and potentially helpful is to recognize that the highlighted importance of public outreach and education is consistent with the endorsed virtue-ethical view.

According to the view, good environmental interventions promote capacities for living healthier lives by protecting or enhancing environmental resources. Yet, even better environmental interventions would amplify that by garnering greater support and promotion of such interventions. Of course, this is also compatible with the above analysis of the AC cases. So too, public outreach and education can help to enhance well-being in a place by making one more a part of it. Recall, in the endorsed view, "part of what it is to live a good life is to contribute in the right way to ongoing social-ecological functioning of the systems to which we belong." Being educated about one's connections, potential functional roles, and possible active stewardship roles can serve to promote and enhance one's senses of belonging and functional contribution. Moreover, insofar as protecting and promoting the well-being of many forms of nonhuman nature can be partially constitutive of our biological well-being, this protection and promotion can enrich our lives through exercising virtues of humility for anthropogenic degradation, and can contribute to our psychological well-being. Public outreach and education can only serve to promote these goods and is therefore itself a public good.

Novel Ecosystem Engineering

"Novel ecosystem" (NE) engineering projects highlight another important dimension of restoration, and one where many other potential key virtues lie. This is the recognition of functional integration of the historical presence of humans and the functional legacies of that presence in a place. For example, without species and hydrologic conditions that can be reestablished, restoration cannot aim toward historical functional and compositional states. This is because natural functions and features that were once present in a place, before humans inhabited it, cannot be reestablished if the decades of human presence have degraded or completely displaced the natural resources necessary for those historical conditions and features. Such cases instead sometimes call for introductions of NEs for which trajectories and management strategies are not yet well defined.[54]

Some Considerations about NEs Considered

A straightforward definition of NEs is given by Hobbs et al., who say:

> A novel ecosystem is a system of abiotic, biotic and social components (and their interactions) that, by virtue of human influence, differ from those that prevailed historically, having a tendency to self-organize and manifest novel qualities without intensive human management. Novel ecosystems are distinguished . . . [by] practical limitations (a combination of ecological, environmental and social thresholds) on the recovery of historical qualities.[55]

Although there is disagreement about how to pick appropriate historical goal conditions, traditional restoration strategies have taken historical ecosystem states as the goal state for restoration projects. Some recent works emphasize that it is increasingly impractical, and even dangerous, to endorse such traditional restoration goals, since unprecedented changes due to anthropogenic climate change and other influences have completely changed the historical trajectories and patterns in most natural systems.[56]

Desjardins, Donhauser, and Barker critically evaluate worries that "promoting NEs implies the acceptance of narrow service-based goals that will leave our management systems vulnerable to abuses leading to ecological degradation."[57] They identify two motivations for this worry, what they call the "anything goes" worry, that are worth considering.

One version of this worry is that the goal of NEs will be to maximize goods and services to humans—the worry being that NEs would treat natural areas merely as means of maximizing and exploiting "natural capital" for human populations, driven by preferences expressed through commercial markets. These include, for example, potentially realizing NEs that maximize food production, water quality, timber and fiber, climate regulation, flood control, and various recreational and aesthetic features. The worry is that promoting NEs implies accepting the pursuit of NEs in whatever form any person or entity might find profitable. This constitutes a worry because many see anthropocentric value schemas as problematic since they've led to most environmental problems in the first place.[58] A second version of the worry is that supporting NEs entails potentially embracing novel ecological functionality that is valuable to some target population, community, or ecological network even though it may not benefit humans. This version of the worry centers on concerns that NEs constitute human actions that may potentially harm naturally occurring ecosystems or biodiversity.

I mention these variations of the worry with NEs here primarily to emphasize the complexity and nuance of the relevant considerations. The virtue-ethical position I have endorsed already provides resources for defusing variations of the "anything goes" worry, because the view suggests numerous counts on which any restoration strategy, and thus an NE, could be more or less functionally good.[59]

Cases

In contrast to the case discussed in an earlier section, where the project was oriented toward reinstating historical wetland conditions, some strategies for improving the ecological stability and habitat in wetland watersheds like those found throughout Ohio can lead to the establishment of NEs and networks of NEs. One such strategy is to create *new* streams and tributaries to reconnect parts of watersheds historically disconnected through channelization efforts to divert water from farmland and other property.[60] Notably, as with historical wetland restoration, such efforts function primarily to change and improve the hydrology of a place. Research shows that surface water and groundwater exchange between rivers and their floodplains are a key factor that influences the functionality and health of adjacent riparian ecosystems.[61]

Huang et al. document an effort to design a stream on the Ohio State University, Marion campus—an effort that necessarily entails promoting NE propagation since it is an effort to create new hydrologic connections. Their study is a design and feasibility analysis, rather than a study of the completed project. Still, the outcomes and conclusions are quite compelling and clear. The creation of such a stream is an effective and sustainable "backyard" solution that can drastically improve ecological functionality, water quality, aquatic and terrestrial habitat, and biodiversity over a large area (in this case roughly 10 hectares). Such an effort is also very cheap—in their estimate, only $100,000 over five years. According to their analysis the creek, along with its many beneficial functions, requires very limited (almost no) continued maintenance or intervention after the initial installation. As they put it, "the creek and wetland will maintain and sustain natural status by themselves."[62] Stream creation thus represents one strategy for promoting beneficial NE development that is practical in many ways, sustainable, and can provide many benefits to human and nonhuman populations in an area.

Another "restoration" strategy that can lead to the development of NEs is what Haram et al. describe as using non-native species as "introduced ecosystems engineers." Introducing non-native species, or simply allowing non-native species to establish naturally, and the resulting influence on ecological composition and functionality, is a form of NE engineering simply because it entails development away from historical ecological conditions. Notably, Haram et al. do not necessarily endorse such efforts, but consider both positive and negative tradeoffs that such efforts can have on relevant ecosystems, and the relevant human and nonhuman populations inhabiting such an ecosystem. In their own summary, their project "illuminates the need to account for species identity, individual behavior, and scale when predicting the impacts of [non-native] . . . species on native communities."[63] That said, however, they do identify numerous benefits that this strategy for NE development has had in certain cases. Introduced ecosystem engineers can have positive functional effects on native ecological community members by changing and generating new habitat.[64] This is why Haram et al. stress the need to look at the impacts of the development of NEs due to introduced ecosystem engineers on *entire ecological communities*, because their relative benefits can often only be seen at that level of consideration. For instance, though not in Ohio, these researchers recount that introduction of herbaceous forest understory in Appalachia by Japanese stilt grass (*Microstegium vinineum*) has reduced arthropod population density. However, it also resulted in increased habitat availability for predatory spiders, which can in turn have positive impacts on the well-being capacities of amphibians and other populations.

One may wonder why favoring spiders and a certain kind of grass over other arthropods is *good* according to the virtue-ethical perspective. It is an instance of an example of using a hands-off approach to letting NEs develop, which in the cited case study and context was good for local spiders and amphibians and birds. In that case, the effort was functionally better for them on the whole than not letting the non-native species spread. The more general point for our purposes is that NE propagation due to introduced ecosystem engineers can have potentially positive and wide-reaching community-level effects on the well-being of numerous species. It thus represents another NE engineering strategy that may be worth considering in ecological contexts where doing so would likely promote the well-being interests of human and nonhuman species in those particular contexts.

A third potential strategy for promoting the development of NEs is rewilding defunct farmland in a manner similar to what has been discussed above, and again without a historical ecological goal state as the aim.[65] Beth Middleton studied the composition of seed banks of farmland in the Southeast, including some in Ohio, to determine their viability for restoring forested wetlands in the manner discussed above. The challenges highlighted in that study underscore the potential for pursuing NE engineering efforts where historical restorations are very difficult or impossible. In sum, Middleton's study stresses that converting farmland back into forested wetland with functional properties including hydrological control and connectivity similar to historical systems depends heavily on the recruitment and development potential of seeds.[66] Yet, species that are most viable under current and future conditions are very unlikely to be of the same kinds as historical species assemblages in areas like Ohio. This is because environmental conditions change across time, especially when humans inhabit and reengineer the ecological, geomorphological, and hydrological functionality of a place.[67] Likewise, Middleton notes that restored floodplain forests in the eastern United States in general do not resemble those that were historically on the relevant land.[68] So, the takeaway is simple: sometimes promotion of NEs is an option for restoring certain beneficial ecological and other natural functions where historical restorations are not feasible, as is the case in numerous potential and actual farmland rewilding efforts in Ohio. Of course, this is because NEs have many more benefits than non-use defunct farmland even if they are not historical. More generally, ecological functions and their potential values can vary independently of historical functionality, especially when historical functionality has been completely destroyed.

Finally, and perhaps unexpectedly, land abandonment is another "strategy" for promoting the development of potentially beneficial NEs. Munroe et al. examine the "alternative trajectories" of cases of land abandonment, including cases in Ohio that can lead to the establishment of functionally valuable NEs. They spotlight a case in southeastern Ohio, in the northern Appalachian region, to illuminate how complex the social and environmental aspects of NEs due to land abandonment can be. There they documented the development of an NE on former strip-mining and agricultural sites from 1955 to 2008, showing that due to its own unassisted growth and development, the former mining site is now completely reforested and the former agricultural sites are now partially reforested and sprinkled with adjacent residential properties.[69] That case alone, and just my brief mention of it, suffices to make the point that land abandonment can, and has, led to the development of functionally valuable NEs. And so land abandonment too represents a potential NE engineering strategy that may be worth considering in contexts where it could enhance and promote the well-being potential and well-being interests of human or nonhuman species.

Discussion

There are two important normative considerations to be added to the discussion of NEs, as evaluated through the lens of the endorsed virtue-ethical position. First, NEs provide means of finding adaptive responses in situations where traditional restoration strategies are unviable, impractical, or impossible due to large-scale and local anthropogenic environmental changes. Accordingly, it is important to think about ecological goods and human and nonhuman well-being interests as adaptive as well. As Thompson argues, doing good in ecological restoration entails identifying valuable novel ecological goods and novel forms of human and nonhuman well-being that may arise as our environments continue to change due to anthropogenic climate change.[70] In other words, climate change "raises the salience of virtues related to openness and accommodation" associated with restoration, "weakens the justification for historical fidelity," and "raises the salience of reconciliation as a virtue" associated with climate change response and restoration.[71] In other words, climate change often makes historical fidelity an infeasible practical target, but at the same time makes finding restoration and management practices that promote sustainable relationships between human and nonhuman populations and their environments more crucial. Accordingly, appreciating the potentials of NEs and "learning how to harness their benefits" is something we now must strive for in considering restoration strategies.[72]

Second, it is important to recognize that more virtuous restoration practices must engage social components in yet another way: they should be "guided by a serious commitment to *facilitating access* to ecological goods."[73] This is consistent with Aristotle's original virtue-ethical conception of realizing well-being capacities. According to Aristotle, realizing one's capacities for well-being requires having both an internal functional capacity and practical ability, and so access to the means of realizing the relevant capacities in situ.[74] Accordingly, no matter how they incorporate historical or novel ecological features, ecological restoration initiatives can be evaluated as better or worse on the basis of how well they provide functional ecological goods and also access to the particular ecological goods of any human or nonhuman individual's local environment. This is context-dependent. For the good life on the Ohio shore of Lake Erie surely differs in the functional particulars from the good life of human individuals in Arizona or Vermont at any given time.[75]

Conclusion

I have examined restoration strategies including farmland restoration, assisted colonization of predators, and "novel ecosystem" engineering initiatives in and around wetlands and disconnected watersheds in Ohio. Along the way, I highlighted positive and negative tradeoffs of each strategy, while focusing on the positive takeaways in order to suggest generalizable insights. I have also outlined a virtue-ethical evaluative framework, and have discussed the implications of the view for how we might usefully evaluate each of the discussed sorts of restoration strategies.

As I said at the outset, the endorsed virtue-ethical perspective also helps deal with a tension between anthropocentric values that often motivate restoration efforts and non-anthropocentric values that many point out must be preserved to effectively and sustainably address overarching issues like global climate change and environmental justice. This is because, as we have seen, the virtue-ethical perspective focuses on enhancing and maximizing the well-being interests and potentials of human and nonhuman nature in each particular situation and context. The perspective therefore conceives of restoration and engineering aims as adaptive and responsive to whatever contextual features may promote overall, holistic well-being. And so these aims are responsive to fulfilling well-being capacities and interests relative to contingencies in each particular context. The discussion above has shown that this makes the virtue-ethical perspective more practically useful than alternative—that is, utilitarian and deontological—views.

Of course, my discussion has not been exhaustive or comprehensive. I have tried to illuminate key aspects of restoration initiatives and practices that should be given attention in the hope of realizing more effective and ethical restorations, and outlined the main tenets of a perspective that is useful for evaluating such initiatives. This discussion of the strategies and the virtue-ethical view should be helpful to others interested in environmental protection and restoration in Ohio, other parts of the Midwest, and beyond.

NOTES

1. See, for example, Katie McShane, "Anthropocentrism vs. Nonanthropocentrism: Why Should We Care?" *Environmental Values* 16, no. 2 (2007): 169–185.

2. Eric Desjardins, Justin Donhauser, and Gillian Barker, "Ecological Historicity, Novelty and Functionality in the Anthropocene," *Environmental Values* 28, no. 3 (2019): 275–303.

3. Desjardins et al., "Ecological Historicity," 296.

4. Holmes Rolston, "Environmental Virtue Ethics: Half the Truth but Dangerous as a Whole," in *Environmental Virtue Ethics*, ed. Ronald Sandler and Phil Cafaro (Lanham, MD: Rowman and Littlefield, 2005), 65.

5. J. M. Cooper, "Aristotle on the Forms of Friendship," *Review of Metaphysics* 30, no. 4 (1977): 619–648, at 625, 627.

6. Desjardins et al., "Ecological Historicity," 295.

7. Kenneth Shockley, "Sourcing Sustainability in a Time of Climate Change: Environmental Values," *Environmental Values* 23, no. 2 (2014): 209–210.

8. Allen Thompson, "The Virtue of Responsibility for the Global Climate," in *Ethical Adaptation to Climate Change: Human Virtues of the Future*, ed. Allen Thompson and J. Bendik-Keymer (Cambridge, MA: MIT Press, 2012), 214.

9. Justin Donhauser, "Environmental Robot Virtues and Ecological Justice," *Journal of Human Rights and the Environment* 10, no. 2 (2019): 176–192.

10. Cf. Ian A. Smith, *The Intrinsic Value of Endangered Species* (New York: Routledge, 2016), chap. 4.

11. William Fitzpatrick, "Valuing Nature Non-instrumentally," *Journal of Value Inquiry* 38, no. 3 (2004): 315–332. See also Smith, *Intrinsic Value*, 118–120.

12. See Smith, *Intrinsic Value*, chap. 7.

13. Donhauser, "Environmental Robot Virtues."

14. D. Schlosberg, *Defining Environmental Justice: Theories, Movements, and Nature* (Oxford: Oxford University Press, 2007), 142.

15. Schlosberg, *Defining Environmental Justice*, 143. Schlosberg prefers to use "capabilities" language; I've changed that to "capacities" simply to maintain consistent language throughout this discussion.

16. Shannon Vallor, *Technology and the Virtues* (New York: Oxford University Press, 2016).

17. Donhauser, "Environmental Robot Virtues," 184.

18. Donhauser, "Environmental Robot Virtues," 184; cf. Vallor, *Technology and the Virtues*, sect. 2.3.

19. Vallor, *Technology and the Virtues*, 25.

20. Donhauser, "Ecological Historicity," 294. This is not to suggest that all utilitarian views are anthropocentric; they are not. Indeed, there are well-known animal welfarist utilitarian views like Peter Singer's, which builds upon Jeremy Bentham's view. See Singer, *Animal Liberation: The Definitive Classic of the Animal Movement*, updated ed. (New York: HarperCollins, 2009).

21. Rolston, "Environmental Virtue Ethics," 65.

22. Desjardins et al., "Ecological Historicity," 295; see also Vallor, *Technology and the Virtues*, 26, 39.

23. See D. Imhoff, *The CAFO Reader: The Tragedy of Industrial Animal Factories* (Los Angeles: Watershed Media, 2010).

24. Julia Lenhardt and Yelena Ogneva-Himmelberger, "Environmental Injustice in the Spatial Distribution of Concentrated Animal Feeding Operations in Ohio," *Environmental Justice* 6, no. 4 (2013): 133–139; J. Barrett, "Hogging the Air: CAFO Emissions Reach into Schools," *Environmental Health Perspectives* 114, no. 4 (2006): A241; M. Mirabelli et al., "Asthma Symptoms among Adolescents Who Attend Public Schools That Are Located near Confined Swine Feeding Operations," *Pediatrics* 118, no. 1 (2006): 2005–2812.

25. Lenhardt and Ogneva-Himmelberger, "Environmental Injustice," 138.

26. Mark Eisler et al., "Agriculture: Steps to Sustainable Livestock," *Nature News* 507, no. 7490 (2014): 32. Paul B. Thompson and Alessandro Nardone, "Sustainable Livestock Production: Methodological and Ethical Challenges," *Livestock Production Science* 61, nos. 2–3 (1999): 111–119.

27. Christian Lenhart and Peter C. Lenhart, "Restoration of Wetland and Prairie on Farmland in the Former Great Black Swamp of Ohio, USA," *Ecological Restoration* 32, no. 4 (2014): 441.

28. Lenhart and Lenhart, "Restoration of Wetland and Prairie," 446.

29. Donhauser offers complementary discussions of the difference and relation between ecological networks and system. See Justin Donhauser, "Making Ecological Values Make Sense: Toward More Operationalizable Ecological Legislation," *Ethics and the Environment* 21, no. 2 (2016): 1–25; Justin Donhauser, "Theoretical Ecology as Etiological from the Start," *Studies in History and Philosophy of Science Part C: Studies in History and Philosophy of Biological and Biomedical Sciences* 60 (2016): 67–76.

30. Lenhart and Lenhart, "Restoration of Wetland and Prairie," 446.

31. William J. Mitsch, "Solving Lake Erie's Harmful Algal Blooms by Restoring the Great Black Swamp in Ohio," *Ecological Engineering of Sustainable Landscapes* 108 (November 1, 2017): 406–413.

32. Lenhart and Lenhart, "Restoration of Wetland and Prairie," 447.

33. See, for example, Mitsch, "Solving Lake Erie's Harmful Algal Blooms."

34. This is not to say that we should judge what interventions are good according to some ecological theory of functionality and functional integration necessarily. Rather, it is to say that restoration and environmental engineering efforts can be judged according to the degree to which the well-being and well-being interests of human and nonhuman species in a place are served by their being more or less causally interrelated.

35. Donhauser, "Environmental Robot Virtues."

36. See, for example, D. H. Getches and D. N. Pellow, "Beyond 'Traditional' Environmental Justice," in *Justice and Natural Resources: Concepts, Strategies, and Applications*, ed. Kathryn M. Mutz, Gary C. Bryner, and Douglas S. Kenney (Washington, DC: Island Press, 2002).

37. Schlosberg, *Defining Environmental Justice*, chaps. 5 and 6. See also Donhauser, "Environmental Robot Virtues" (sect. 4) for analysis of the distinction between environmental and ecological justice.

38. Desjardins et al., "Ecological Historicity."

39. M. L. Hunter, "Climate Change and Moving Species: Furthering the Debate on Assisted Colonization," *Conservation Biology* 21 (2007): 1356–1358; O. Hoegh-Guldberg et al., "Assisted Colonization and Rapid Climate Change," *Science* 321 (2008): 345–346; J. S. McLachlan, J. J. Hellmann, and M. W. Schwartz, "A Framework for Debate of Assisted Migration in an Era of Climate Change," *Conservation Biology* 21 (2007): 297–302; D. M. Richardson et al., "Multidimensional Evaluation of Managed Relocation," *Proceedings of the National Academy of Sciences* 106 (2009): 9721–9724; P. J. Seddon, "From Reintroduction to Assisted Colonization: Moving along the Conservation Translocation Spectrum," *Restoration Ecology* 18 (2010): 796–802.

40. This point of clarification is due to the reviewers and editors; I thank them for it.

41. J. Travis, "Climate Change and Habitat Destruction: A Deadly Anthropogenic Cocktail," *Proceedings of the Royal Society B: Biological Sciences* 270 (2003): 467–473.

42. Ronald Sandler, "The Value of Species and the Ethical Foundations of Assisted Colonization," *Conservation Biology* 24 (2010): 424–443.

43. Thomas L. Serfass et al., "Otters and Anglers Can Share the Stream! The Role of Social Science in Dissuading Negative Messaging about Reintroduced Predators," *Human Dimensions of Wildlife* 19, no. 6 (2014): 532–544.

44. See H. Kruuk, *Otters: Ecology, Behaviour and Conservation* (New York: Oxford University Press, 2006); W. E. Melquist, P. J. Polechla Jr., and D. Toweill, "River Otter," in *Wild Mammals of North America: Biology, Management, and Conservation*, ed. G. A. Feldhamer, B. C. Thompson, and J. A. Chapman (Baltimore, MD: Johns Hopkins University Press,

2003).

45. Thomas L. Serfass et al., "Otters and Anglers," 533. Cf. T. L. Goedeke, "Devils, Angels or Animals: The Social Construction of Otters in Conflict over Management," in *Mad about Wildlife: Looking at Social Conflict over Wildlife*, ed. A. Herda-Rapp and T. L. Goedeke (Boston: Brill, 2005), 25–50; D. Hamilton, "From Near Zero to Fifteen Thousand—in 20 Years! Missouri's River Otter Saga," *River Otter Journal* 15 (2006): 1–12.

46. Steve Pollick, "Otters Everywhere," *Ohio Sportsman* (blog), December 5, 2004, https://www.ohiosportsman.com/threads/ohio-otters-everywhere.5827/.

47. "Ohio River Otter Captured in Putnam County," Outdoor Hub, January 5, 2012, https://www.outdoorhub.com/pr/2012/01/05/ohio-river-otter-captured-in-putnam-county/.

48. See Thomas L. Serfass et al., "River Otters in Pennsylvania, USA: Lessons for Predator Reintroduction," in *Proceedings of the European Otter Conference "Return of the Otter in Europe: Where and How,"* ed. J. W. H. Conroy et al. (Broadford, UK: International Otter Survival Fund, 2003).

49. Thomas L. Serfass, "Reintroduction of Woodrats: Concepts and Applications," in *The Allegheny Woodrat: Ecology, Conservation, and Management of a Declining Species*, ed. John Peles and Janet Wright (New York: Springer-Verlag, 2008), 170. See also Walter J. Schlie, "Reintroduction of the Allegheny Woodrat (*Neotoma floridana magister*) to Neotoma Valley, Hocking County, Ohio" (PhD diss., The Ohio State University, 1985); Walter J. Schlie and T. A. Bookhout, "Reintroduction of Allegheny Woodrats to Hocking County, Ohio," *Ohio Journal of Science* 85, no. 2 (1985): 92.

50. Serfass, "Reintroduction of Woodrats," 181. Cf. Devra G. Kleiman, "Reintroduction of Captive Mammals for Conservation," *BioScience* 39, no. 3 (1989): 152–161; D. G. Kleiman et al., eds., *Wild Mammals in Captivity* (Chicago: University of Chicago Press, 1996).

51. See Steven H. Fritts et al., "Planning and Implementing a Reintroduction of Wolves to Yellowstone National Park and Central Idaho," *Restoration Ecology* 5, no. 1 (1997): 7–27.

52. Thomas L. Serfass et al., "Genetic Variation among Populations of River Otters in North America: Considerations for Reintroduction Projects," *Journal of Mammalogy* 79, no. 3 (August 21, 1998): 736–746. See also Serfass et al., *Proceedings of the European Otter Conference*.

53. Serfass, "Reintroduction of Woodrats," 181.

54. T. R. Seastedt, R. J. Hobbs, and K. N. Suding, "Management of Novel Ecosystems: Are Novel Approaches Required?" *Frontiers in Ecology and the Environment* 6 (2008): 547–553.

55. Richard J. Hobbs, Eric S. Higgs, and Carol M. Hall, "Defining Novel Ecosystems," in

Novel Ecosystems: Intervening in the New Ecological World Order, ed. Richard J. Hobbs, Eric S. Higgs, and Carol M. Hall (Chichester, UK: Wiley-Blackwell, 2013), 58.

56. See, for example, Young D. Choi, "Restoration Ecology to the Future: A Call for New Paradigm," *Restoration Ecology* 15, no. 2 (2007): 351–353; Desjardins et al., "Ecological Historicity"; R. J. Hobbs et al., "Intervention Ecology: Applying Ecological Science in the Twenty-first Century," *BioScience* 61, no. 6 (2011): 442–450; Marion Hourdequin, "Restoration and History in a Changing World: A Case Study in Ethics for the Anthropocene," *Ethics & the Environment* 18, no. 2 (2013): 115–134.

57. Desjardins et al., "Ecological Historicity," 277.

58. Cf. Bryan G. Norton, "Environmental Ethics and Weak Anthropocentrism," *Environmental Ethics* 6, no. 2 (1984): 131–148; Sahotra Sarkar, *Biodiversity and Environmental Philosophy: An Introduction* (Cambridge: Cambridge University Press, 2005).

59. Desjardins et al., "Ecological Historicity," 281; Andrew Light, Allen Thompson, and Eric S. Higgs, "Valuing Novel Ecosystems," in Hobbs, Higgs, and Hall, *Novel Ecosystems*, 257–268.

60. One might object that reconnecting parts of watersheds is not *creating* new streams (that is, streams where there historically were not any). However, I'm explicitly talking about efforts where new streams are created to reconnect watersheds. This is very often necessary to reconnect parts of watersheds because property rights and manmade structures (for example, roads) can make it impossible to reestablish rivers where old riverbeds were. Moreover, the biotic assemblages and/or fluvial geomorphology has often changed in places such that reestablishing old rivers is not possible.

61. Wolfgang J. Junk, Peter B. Bayley, and Richard E. Sparks, "The Flood Pulse Concept in River-Floodplain Systems," *Canadian Special Publication of Fisheries and Aquatic Sciences* 106, no. 1 (1989): 110–127; Klemnent Tockner, Florian Malard, and J. V. Ward, "An Extension of the Flood Pulse Concept," *Hydrological Processes* 14, nos. 16–17 (2000): 2861–2883.

62. Jung Chen Huang, William J. Mitsch, and Li Zhang, "Ecological Restoration Design of a Stream on a College Campus in Central Ohio," *Ecological Engineering* 35, no. 2 (2009): 338.

63. Linsey E. Haram et al., "Mixed Effects of an Introduced Ecosystem Engineer on the Foraging Behavior and Habitat Selection of Predators," *Ecology* 99, no. 12 (2018): 2751–2762.

64. See P. E. Gribben et al., "Positive versus Negative Effects of an Invasive Ecosystem Engineer on Difference Components of a Marine Ecosystem," *Oikos* 122 (2013): 816–824; J. L. DeVore and J. C. Maerz, "Grass Invasion Increases Top-down Pressure

on an Amphibian via Structurally Mediated Effects on an Intraguild Predator," *Ecology* 95 (2014): 1724–1730; J. T. Wright et al., "Engineering or Food? Mechanisms of Facilitation by a Habitat-Forming Invasive Seaweed," *Ecology* 95 (2014): 2699–2706.

65. Some readers may reasonably wonder how creating NEs can be a kind of *rewilding*, since rewilding is often conceived of as returning an area to a pre-human-presence wild state. However, in my view, wildness does not necessarily have anything to do with returning an area to a previous state. The wildness of an area is simply its degree of functional autonomy from human interventions and influence.

66. Beth A. Middleton, "Soil Seed Banks and the Potential Restoration of Forested Wetlands after Farming," *Journal of Applied Ecology* 40, no. 6 (2003): 1025–1034.

67. See Eric Higgs, *Nature by Design: People, Natural Process, and Ecological Restoration* (Cambridge, MA: MIT Press, 2003). Higgs's piece is a useful resource on this point.

68. Middleton, "Soil Seed Banks."

69. Darla K. Munroe et al., "Alternative Trajectories of Land Abandonment: Causes, Consequences and Research Challenges," *Current Opinion in Environmental Sustainability* 5, no. 5 (2013): 471–476.

70. Thompson, "Virtue of Responsibility," 214.

71. Desjardins et al., "Ecological Historicity," 296.

72. Desjardins et al., "Ecological Historicity"; Light, Thompson, and Higgs, "Valuing Novel Ecosystems."

73. Desjardins et al., "Ecological Historicity," 296.

74. Aristotle, *Nicomachean Ethics* 1098a6–13; see also Shockley, "Sourcing Sustainability," 209–210.

75. Desjardins et al., "Ecological Historicity," 297.

SUGGESTED READINGS

Desjardins, Eric, Justin Donhauser, and Gillian Barker. "Ecological Historicity, Novelty and Functionality in the Anthropocene." *Environmental Values* 28, no. 3 (2019): 275–303.

Donhauser, Justin. "Making Ecological Values Make Sense: Toward More Operationalizable Ecological Legislation." *Ethics and the Environment* 21, no. 2 (2016): 1–25.

Fitzpatrick, William J. "Valuing Nature Non-instrumentally." *Journal of Value Inquiry* 38, no. 3 (2004): 315–332.

Higgs, Eric. *Nature by Design: People, Natural Process, and Ecological Restoration.* Cambridge, MA: MIT Press, 2003.

Hunter, M. L. "Climate Change and Moving Species: Furthering the Debate on Assisted Colonization." *Conservation Biology* 21 (2007): 1356–1358.

Light, Andrew, Allen Thompson, and Eric S. Higgs. "Valuing Novel Ecosystems." In *Novel Ecosystems: Intervening in the New Ecological World Order*, edited by Richard J. Hobbs, Eric S. Higgs, and Carol M. Hall, 257–268. Chichester, UK: Wiley-Blackwell, 2013.

Sandler, Ronald. "The Value of Species and the Ethical Foundations of Assisted Colonization." *Conservation Biology* 24 (2010): 424–443.

Schlosberg, D. *Defining Environmental Justice: Theories, Movements, and Nature*. Oxford: Oxford University Press, 2007.

Smith, Ian A. *The Intrinsic Value of Endangered Species*. New York: Routledge, 2016.

Thompson, Allen. "The Virtue of Responsibility for the Global Climate." In *Ethical Adaptation to Climate Change: Human Virtues of the Future*, edited by Allen Thompson and J. Bendik-Keymer, 203–222. Cambridge, MA: MIT Press, 2012.

Are the Next Generation Science Standards for Weather and Climate Indoctrinating?

Matt Ferkany

The climate change problem is hugely complicated and, by its nature, confronts us with many uncertainties. Precisely how much warming will occur given predicted levels of greenhouse gas emissions? What will be the full climatological and environmental consequences of this warming? And how damaging will these consequences be for humans economically or geopolitically?

We might expect, given such complexity and uncertainty, that experts on climate change would often disagree, perhaps vehemently, about many aspects of the problem. Despite these uncertainties, the evidence is that climate scientists agree on certain fundamentals—in particular that climate change is occurring, and human activity has been the primary driver of it in the past one hundred years.[1]

Nevertheless, an ongoing war has been waged, especially in the American Midwest and Plains States, between advocates and critics of adopting updated science standards based on the Next Generation Science Standards (NGSS) in schools.[2] Critics have generally opposed efforts to standardize the educational curriculum across the fifty states. But they have strenuously objected in particular to the NGSS for weather and climate. These standards would update the teaching of weather and climate science to include the thesis that human activity is a major cause of recent climate change (the anthropogenic global climate change thesis, or AGCC thesis). Standards teaching the AGCC thesis are indoctrinating because scientists disagree, the critics claim, about this thesis and it is indoctrinating to teach only one side of an unsettled scientific controversy.[3]

Are the NGSS for weather and climate one-sided and therefore indoctrinating? What is involved in indoctrination anyway? This chapter will build on the notion that indoctrination is the inculcation of unjustified certainty in a belief—dogmatism, in other words—to show that educators teaching the NGSS for weather and climate would not be guilty of indoctrinating their students, or at least, not in the way critics claim. Children have a right to be educated in ways that prepare them to function well and cope in the world they will inherit. According to the best available science, climate change is expected to pose many serious problems for their world. Numerous surveys of expert scientists find that they almost unanimously agree that human activity is causing observed climate change, that is, changes in earth's global and regional weather and climate systems due to a long-term trend of rising global surface temperatures.[4] Efforts to deny that this is not the case are not credible. Therefore, adherence to the consensus view of the AGCC thesis would not be dogmatic; that is, it would be justified. Therefore, it is not troublingly one-sided not to teach it, and not indoctrinating. If anything, the NGSS risk being indoctrinating in a different way: they potentially inculcate highly technocratic and environmentally conservative attitudes toward the causes and solutions to deeply moral issues like climate change.

The discussion of this chapter unfolds as follows. The first section discusses science education, the aims of the NGSS, and the current state of NGSS implementation. The second discusses the moral and ideological culture of the Midwest and resistance to the NGSS for weather and climate. The third makes an argument for teaching the NGSS climate standards and discusses the expected impacts of climate change on the Midwest. The next section discusses the concepts of indoctrination and the indoctrination charge. The fifth argues that the NGSS are not indoctrinating, even on a weak standard of indoctrination, and the sixth advances additional argumentation against the critics. The concluding section further discusses the current state of NGSS implementation and some genuinely troubling moral aspects of the standards.

The NGSS, the IPCC, and the Nature of Science

Science education historically has been accused of suffering serious organizational and messaging problems. At the K–12 level particularly, science (including biological and environmental science) has often been taught as a vast body of received knowledge that students must simply absorb. This message conveys a false image of science and scientific knowledge: Scientific knowledge is certain and unchanging rather than probabilistic and changeable; its methods are mechanical and inevitable; the world imposes itself on us through data that need no interpretation; and new results can be uncovered by any sufficiently dedicated and ingenious researcher working alone in a lab.

This is untrue to the nature of scientific knowledge and the methods and values of science.[5] Aspects of science—such as cell theory, evolutionary biology, or the kinetic molecular theory of gases—are settled consensus science; they are unlikely to be overturned any time soon by new evidence. But science does involve a peculiarly skeptical and social set of intellectual methods and values. The myth of the brilliant scientist revolutionizing science alone in a lab is just that: a myth. Scientists are just ordinary human beings. They have limited knowledge of the world. The knowledge they do have is inevitably colored by their peculiar interests, perspectives, and indeed biases.

Scientists collectively check their interpretations and control for their biases by faithfully sharing the results of their investigations and subjecting them to the scrutiny of other scientists. In academic science, publishing in double-blind peer-review journals—in which the identities of the author(s) and the peer reviewers are hidden from each other—is an important mechanism of this collective self-checking. Replication of prior research results; participation in professional organizations, such as the National Academy of Sciences; and presenting work at conferences are others. These mechanisms are not perfect. Because (often) there are only so many experts in a given field, the veil of blind review is sometimes thin. The intellectual culture of a professional organization can also be captured by an aging scientific paradigm and thus be slow to incorporate new ideas, approaches, or people. However, the corrective to these shortcomings is not, say, abandoning them for their opposites. It is better peer review, or more inclusivity, all while upholding common standards of rigor for research.

Though it does not conduct original research, the Intergovernmental Panel on Climate Change (IPCC) is regarded as the world's leading source of climate information as it incorporates multiple elements of the ideal of collective self-correction. Founded by the United Nations Environment Program and the World Meteorological Organization in 1988, the mission of the IPCC is "to prepare a comprehensive review and recommendations with respect to the state of knowledge of the science of climate change; the social and economic impact of climate change, and potential response strategies and elements for inclusion in a possible future international convention on climate."[6] The IPCC undertakes this mission through a massively collaborative effort consisting of 159 member nations and hundreds of administrators and scientists who write or review the reports on a volunteer basis. Authors and reviewers are experts working in dozens of different fields at academic institutions across the globe in both developed and developing nations.

The process of collective bias correction is expected over time to yield new insights that can change our understanding, sometimes resulting in significant changes to our conceptual schemes. For this reason, prominent contemporary accounts of the nature of science for science education highlight the probabilistic nature of scientific knowledge, as well as the social and creative (in addition to the analytic or quantitative) skills needed in the practice of science. Research into the kinds of people who are excellent practitioners of science, for example, emphasizes characteristics like curiosity, open-mindedness, trust, and humility to evidence.[7]

The NGSS are the result of a movement to transform the teaching of science to better communicate the nature of science itself. The point is to improve not just the content of K-12 science information, but the image and understanding of the methods and skills of science that content projects. The NGSS aim to reorient the teaching of science to help students understand the nature of science by teaching them to think like scientists.[8]

Naturally, given the changing nature of scientific knowledge, the NGSS also include updates to specific scientific ideas, including in the study of weather and climate. Whereas older standards documents may mention basics of climate and climate change—such as that earth's climate system relies on heat from solar radiation—the NGSS are explicit in certain passages that human activity is a cause of climate change. At the high school level (designation "HS"), the NGSS include the following:

HS-ESS2.D3: Changes in the atmosphere due to human activity have
increased carbon dioxide concentrations and thus affect climate.

HS-ESS2.D4: Current models predict that, although future regional climate
changes will be complex and varied, average global temperatures will
continue to rise. The outcomes predicted by global climate models
strongly depend on the amounts of human-generated greenhouse gases
added to the atmosphere each year and by the ways in which these
gases are absorbed by the ocean and biosphere.

HS-ESS3.D1: Though the magnitudes of human impacts are greater than
they have ever been, so too are human abilities to model, predict, and
manage current and future impacts.[9]

Needless to say, the incorporation of such standards is a significant improvement
upon past weather and climate standards in which the AGCC thesis is absent. The
AGCC thesis is now settled science among the scientists most qualified to evaluate
it, those claiming climatological expertise and having published extensively on
climate change in peer-reviewed journals.[10]

Likewise, the overall intent of the NGSS (to better communicate the nature of
science and to teach students how to think like scientists) is salutary for a society
in which many of the biggest problems—environmental problems especially—are
scientifically complex. The germ theory of disease is settled science. However new
diseases present new puzzles and the solutions to them may sometimes be solved
only through trial and error. This is normal science. If citizens misunderstand the
nature of environmental science, say, they are liable to make poor decisions, especially concerning which sources of scientific information to trust and which not
to trust. Recent updates to Centers for Disease Control (CDC) recommendations
for public health measures relating to the COVID-19 pandemic have been a case
in point. Public mistrust of the CDC spiked when they first denied, then later
affirmed, that mask wearing is effective for reducing the spread of COVD-19.

Climate Change Opinion in the Midwest

In the American moral imagination, the Midwest occupies a fairly distinctive place.
Occupying a cultural/ideological area somewhat wider than the official geopolitical
map, the Midwest is the moral center of the country, its "heartland." In this

ideological space, the Midwest is geographically different from, but fundamentally cut from the same moral cloth as, the more northern states of the South—such as Virginia, widely considered a Southern state likely because of its place in the Civil War—and the Plains States running up from Oklahoma to Montana.[11]

This American heartland is one relatively conservative part of the country, one where its corn and wheat are grown, its cars are made, and its cattle graze. It is rich in natural treasures, such as the Great Lakes, the foothills of the Appalachian Mountains, and the prairies. But notably it is not a part of the country widely associated with deep concern for environmental causes. That honor largely belongs to the pioneer West and the liberal Pacific and Northeastern coasts. Among the central values that spring to mind in thinking about the American Midwest, environmentalism is not one of them.

The relative environmental conservatism of the Midwest is reflected in how convinced—or rather unconvinced—Midwesterners are about the reality and severity of anthropogenic climate change relative to the scientific consensus.[12] Opinion about climate change in the Midwest is variable, though it largely reflects the problems of climate denialism that exist in pockets of the United States as a whole. Many Midwesterners are very concerned about climate change whereas others are very unconcerned; 63 percent of Illinoisans are worried while only 54 percent in neighboring Indiana are worried. This climate of opinion is quite different, however, from that among scientists and in some other regions of the country where concern is more uniformly high. Citizens of coastal states, including conservative states like North Carolina (62 percent), are overall more worried than many of their Midwestern counterparts. Many people in the Midwest deny that humans are causing climate change and are less concerned that anything be done about it. While 55 percent of Americans agree that most scientists think that global warming is happening, in many Midwestern states agreement falls at or below 50 percent (Indiana, Kansas, North Dakota, South Dakota, Nebraska).

It should little surprise us that the "moral" Midwest is home to some of the most vocal and powerful American advocates of climate change denialism. Notoriously, in February 2015, the senator from Oklahoma James Inhofe went before Congress—armed with a snowball collected from outside the Capitol steps—to argue that it was "unseasonably cold" outside in Washington, DC, and therefore climate change must be a hoax. (The senator also published a book in 2012 entitled *The Greatest Hoax*.) The bizarreness of such an argument in the middle of February—also known as winter—boggles the mind, never mind that the previous year (2014) had been (another) hottest on record, that is, until that point. Born in Wichita, Kansas, where the family business still operates, Midwesterners Charles and David Koch have also been leading funders of denialist think tanks. While the linkages are unclear in this case, this includes funds given to the Illinois-based Heartland Institute, a leading think tank promoting denialist arguments in science education in schools.[13]

In 2017, Heartland circulated 350,000 copies of their book *Why Scientists Disagree about Global Warming: The NIPCC Report on Scientific Consensus* to K–12 science teachers across the country.[14] The authorship presents the work as the product of an apparently international collaborative effort—the Nongovernmental International Panel on Climate Change (NIPCC)—akin to the massively collaborative reports of the IPCC. In fact, the "NIPCC" consists of a non-academically employed geographer (Idso), one retired paleontologist (Carter, now deceased), and one retired physicist (Singer, now deceased), funded by a partisan think tank.[15] The stated mission of the "NIPCC" is "to discover, develop, and promote free-market solutions to social and economic problems."[16]

Such tactics have seemingly born fruit. While critics of the NGSS weather and climate standards may in the end lose the war, they may have won crucial battles. Teachers remain either hesitant to address climate change in their classes, or misinformed and unprepared to do so.[17]

As of the time of writing, and in some states, teachers will have to wait a bit longer before they will have the backing of state-mandated standards to teach the consensus view of climate change without fear of reprisal. The NGSS will not be implemented in Oklahoma until the 2021 school year. In Minnesota, updated standards for weather and climate will not be implemented until 2023. Meanwhile, the IPCC estimates that an 80 percent reduction in carbon emissions in the United States by 2050 will be needed to prevent the worst effects of climate change.[18] Children born this year (2020) who receive an appropriate education in weather and climate science will not be of voting age until 2038.

It is hardly clear that states that have incorporated climate science standards inspired by the NGSS framework have done so in a one-sided way. Indeed, if anything, it looks in many cases as though the process for standards adoption resulted in significant dilution of the standards relating to anthropogenic climate change, which is often left altogether implicit in the standards or rarely explicit.[19] We will return to the actual state of NGSS implementation in the concluding section.

But suppose that states go ahead and adopt every standard of the NGSS requiring teaching of the AGCC thesis. Would they thereby be guilty of indoctrinating schoolchildren?

Probable Impacts of Climate Change on the Midwest

How we answer this question will partly depend on what indoctrination amounts to, an issue I will address in the next section. But surely it depends in part on how concerned (or unconcerned) Midwesterners should be about climate change in the first place. Educators have a duty to inform children of the (probable) conditions of their world, particularly if those conditions are likely to pose problems they will have to manage.[20] Climate change is expected to pose serious problems future generations will have to manage. Thus, educators have a duty to inform children of these problems and enable them to think critically about how to deal with them. This is so even though there are some uncertainties concerning the precise nature and severity of the problems, and corresponding controversy over how to deal with them.

Any time we are dealing with the future, there is sure to be some degree of uncertainty. This does not justify silence or neutrality about such matters in the school curriculum. Suppose NASA and other international space agencies and scientists predict that a meteor is on course to strike the earth in seventy-five years. The precise place the meteor will strike and the full extent of the harms that will result are not known. However, given the meteor's size, they estimate with 95 percent confidence that it will strike the earth with the force of several 15,000 kiloton nuclear weapons (each a thousand times more powerful than the 15-kiloton bomb dropped on Hiroshima). It is also known that—apart from the immediate harms in the vicinity of the impact—this event will have certain kinds of broad and systemic consequences for earth's systems, such as atmospheric cooling and changed weather patterns, shorter growing seasons and increased food scarcity, reduced air quality and higher rates of asthma, and many other similar consequences. Suppose further that with careful planning and economic investment, scientists are 95 percent confident that the meteor could be broken up, if not diverted altogether, thus greatly reducing the harm of impact, as well as mitigating the social and economic pain of coping with unavoidable harms. However, to ensure this outcome, action must be taken sooner rather than later, even though waiting would increase scientists' ability to more precisely estimate the location of impact and full extent of the resulting harm.

I hope it is obvious that it would be wrong of educators to avoid this issue in their teaching. More controversially—though hopefully not much more—it would be wrong to teach children that they need not worry about the problem at all either because there is uncertainty about aspects of it (for example, precisely where impact will occur), or because of skeptical doubts (for example, that there is a problem at all) aired by think-tank scientists or "small-government" politicians. Children in such a world would have a right to know the best available information about this problem, including worst-case scenario information, and a right to think critically about what could or should be done given that information.

Climate change is expected to pose many severe and systemic problems that will threaten human life and health, such as increased heat waves, severe weather events, drought and flooding, and many further knock-on effects like reduced air quality and more air- and water-borne diseases (not to mention migration, property damage, increased food and energy prices, and so on).[21] There are some uncertainties about the precise nature or severity of many consequences of climate change. However, given what is already known from environmental sciences (such as those about ecology, or the cycling of matter like carbon and nitrogen through atmospheric and biospheric systems), the kinds of impacts we can expect in different regions are well understood. No region will be left untouched.

Midwesterners are fortunate enough to live in one of the world's most hospitable regions, one having a long growing season and relatively livable seasonal temperatures. (However, air temperatures are subject to large seasonal swings, with average summer highs around 85 degrees Fahrenheit and average lows around 15°F!) They might be forgiven for doubting that climate change will have much impact on their lives or the region. Compared to people in states in the South and the coasts (60 percent or more in Texas, California, Florida, and Virginia), fewer Midwesterners believe that climate change will harm people in the United States (between 50 and 55 percent in Indiana, Nebraska, and South Dakota), and even fewer that climate change will harm them personally (roughly 30–35 percent compared to the 43 percent national average).[22]

Still, healthy majorities of Midwesterners believe that climate change will harm future generations of people, even in relatively conservative states such as Indiana (63 percent) and North Dakota (61 percent) (the national average is 71 percent). In fact, Midwesterners have reasons to worry about the probable impacts of climate change in their lifetime. Summarizing NOAA data on the region, Charles Fletcher suggests that they are already being impacted by it:

> Heavy downpours are now twice as frequent as they were a century ago, and both summer and winter precipitation have been above average for the last three decades, the wettest period in a century. The Midwest has experienced both increasing extreme events and long-term trends: two record-breaking floods in the past two decades, a decrease in lake ice (including on the Great Lakes), and increased frequency of large heat waves since the 1980s, which have been more frequent than any time in the past century, other than the Dust Bowl years of the 1930s.[23]

In general, average temperatures in the Midwest are expected to continue rising, resulting in both increased flooding, especially in the north, and increased dry days, causing water scarcity and drought, especially in southern states like Kansas and Oklahoma. For example, Kansas experienced its worst drought in thirty years in 2012 with every one of its 105 counties in emergency drought status.[24] And as of the time of writing (in 2020), the current drought has pushed 74 of those counties into drought emergency, warning, or watch status.[25]

The severity of these changes is expected to vary depending on how quickly and extensively humans reduce greenhouse gas emissions, though some are now unavoidable even under "low-emissions" scenarios. Under low-emissions scenarios, deadly heat waves—like one that killed seven hundred people in Chicago in 1995—are expected "to occur every other year in Chicago by the end of the century; under a high scenario, there would be approximately three such heat waves per year."[26]

Children who are ten years old as of the time of writing (2020) will be seventy by 2080. During the period 2004–2018, individuals in this age range (sixty-five years of age or older) were the most susceptible to heat-related deaths across all age groups, accounting for 39 percent of deaths.[27] The future lifetime consequences of climate change for school-age children in 2020 could be very serious indeed. For this reason, educators arguably have an obligation to educate them about this threat, about its causes, consequences, and possible correctives.

The NGSS direct schools to do precisely that. Thus, for the same reason, it is not at all obvious that the NGSS are indoctrinating. It is difficult to see what is indoctrinating about educating children in ways that prepare them to cope with potential serious threats to their lives and the healthy functioning of their society. It is certainly possible that teaching children the received science of climate change will have costs for them. If, as is unlikely, it should turn out that the received science is substantially mistaken, they may choose, for example, to invest scarce resources in dealing with an illusory problem. This would have the effect of diverting those resources away from other serious and unquestionably real problems, such as global inequality or species mass extinction.

However, even in this case, it would not straightforwardly follow that an NGSS education would have been indoctrinating. There is a difference between indoctrinating students with a belief, on the one hand, and sincerely teaching (as if it were true) something that turns out, for reasons one could not have predicted, to be false. Were there not, any past teaching that turned out to be false would have been indoctrinating. That does not seem to be the case. What is indoctrination anyway? When is teaching indoctrinating?

What Is Indoctrination Anyway? The Indoctrination Charge Elaborated

In everyday thinking, indoctrination (when successful) is a kind of brainwashing. This commonsense notion is helpful in highlighting how indoctrination is morally problematic. We have a substantial interest in being connected with reality. Freedom of thought, expression, and choice are also fundamental human rights. Teaching that is indoctrinating violates these interests and rights by exercising undue influence on the mind or beliefs of the student.

To say that indoctrination is brainwashing is to speak metaphorically. What literally does it mean? Precisely what kind or degree of influence over a learner's thought is morally problematic in the way involved in indoctrination? To be sure, teaching that is perfectly legitimate can—and hopefully will—have a lasting impact on students' beliefs and values. Some things, once taught, should not be questioned again. Good teachers should hope that their students leave school with many firm beliefs—for example, that 2 + 2 = 4, or that American democracy is a constitutional system intended to constrain the power of any one branch of government. They should also hope that their students come away with certain firm values as learners and citizens, such as that responsible belief involves being responsive and not indifferent to the evidence for those beliefs, or that voting is a fundamental democratic right and civic duty.

It is clear that indoctrination involves activity exerting undue influence on the mind and beliefs of the learner. Characterizing this influence has been a central source of disagreement about the nature of indoctrination. An influential early view is that indoctrination involves teaching that produces "unshakable" belief in the student.[28] But others have claimed, plausibly enough, that lesser influence on students' beliefs can also be indoctrinating.[29] A teacher may indoctrinate, they claim, by causing students to become relatively closed-minded to alternatives.

In addition to disagreement about how to characterize the effect of indoctrination on the learner's beliefs, scholars have disagreed about whether indoctrination presupposes a malicious intent (that is, a teacher intentionally trying to inculcate a desired belief); about whether it involves any particularly manipulative teaching methods, such as those that control students' feelings without their conscious awareness, or control their reasoning by limiting their access to evidence; and about whether teachers can indoctrinate students with the truth, or only with false beliefs.[30]

For the purposes of this chapter, I will assume the weakest, least stringent possible standard of indoctrination. By "least stringent standard," I mean the standard on which it will be easiest, maybe even common, for teaching to be guilty of indoctrinating. In making this assumption, my aim is not to belittle the wrong of indoctrination nor to claim that indoctrination is in fact common. Rather it is to refute the strongest possible argument on behalf of the critics of the NGSS. If teaching the NGSS for weather and climate would not be indoctrinating on a weak, lax standard, then it would not be indoctrinating on more stringent standards.

I believe that teaching the NGSS for weather and climate would not be indoctrinating on even very weak, easy-to-satisfy standards of what it involves. Therefore, it would not be indoctrinating on more demanding standards. If so, the charge that the NGSS for weather and climate are indoctrinating is false.

What is the weakest, least stringent standard? Indoctrination involves any teaching that leads students to adhere dogmatically to a belief—that is, to adhere to it whatever the evidence (or arguments) for or against it, with the result that it is very difficult to persuade the student to abandon the belief through the presentation of evidence or reasoned argumentation.

This is a weak standard of indoctrination for the following reasons. First, on this standard, teaching can be indoctrinating even if it does not result in "unshakable" belief. It is enough that the learner comes to believe with relative imperviousness (resistance) to relevant evidence, so that it is very difficult to persuade her to change her mind in light of it.

Second, the standard does not require any intentional effort on the part of the teacher to indoctrinate. Certainly, a teacher may indoctrinate by knowingly presenting evidence (for or against) in a biased way, for the purpose of inculcating resistance to it. But a teacher need not try in order to produce such resistance. Such resistance might be produced if, for example, the best available evidence is incomplete or biased in some way that, perhaps for systemic social reasons, makes it difficult even for responsible educators to see. In these sorts of cases, due responsiveness to evidence is blocked by larger social causes that do not lead us to impugn either the intellectual skill or good intentions of individuals, such as when cultural horizons are very narrow or access to disconfirming evidence is very limited. For example, an Amish schoolteacher who has never traveled outside the community or has no access to modern media might indoctrinate children in traditional Amish beliefs in this way; conversely, outsiders to Amish and other communal-agrarian ways of life might indoctrinate children to be hostile to them through sheer ignorance of what they really involve.[31]

Third, this standard does not require any particular method of teaching. Indoctrinating teachers may, or may not, present evidence in ways that manipulate students' feelings, and they may or may not present it in deliberately biased ways, so that students are unable to give a fair hearing to contrary evidence.

Finally, it is possible, in this view, to indoctrinate students with the truth. The worst cases of indoctrination will involve being made to adhere (without due regard for relevant evidence) to what is false. But even those whose beliefs are true can be dogmatic, that is, believe them with a high degree of certainty despite being unable to produce or cite justifying evidence (or to put this differently, despite being able to produce only sham evidence—evidence that is obviously not justifying).

Critics of the NGSS climate standards seem to believe that any teaching that aligns with the standards would be indoctrinating in this sort of way. Teaching to these standards may or may not result in students unshakably convinced that humans are causing climate change. They may simply be irrationally resistant to evidence to the contrary. Teachers who teach to the standards also need not intend to indoctrinate nor deploy any suspect method of teaching. In the critic's view, because scientists actually disagree about the AGCC thesis (so they claim) and this disagreement would not be taught, science teachers would indoctrinate just by faithfully following the standards (so the argument goes) and presenting the evidence that aligns with them as accurately as they can. Because the evidence aligning with the standards is biased, supposedly, the standards are biased; any fair presentation of that evidence would therefore also be biased.

Critics might add to this (cynically?) that it does not even matter whether the AGCC thesis is or is not true. Teaching the weather and climate standards are ironically out of step, they might say, with the motivating purpose and aim of the NGSS movement either way. As discussed above, that movement aims to restructure the teaching of science to highlight the evolving nature of scientific knowledge, the creative and analytic skills of scientists over science content, and the core values of scientific inquiry, such as open-mindedness, humility to evidence, and curiosity. The weather and climate standards are out of step with this purpose, they might say, because they treat the AGCC thesis as settled science. But "true science is never 'settled,' and true scientists are always eager to ask and answer questions. This is plainly the case regarding climate change."[32] If indoctrination is the inculcation of belief unduly resistant to evidence (that is, dogmatism), then teachers who wish to avoid indoctrination should generally encourage in their students a habit of examining both sides of any genuine controversy. This is especially the case with scientific controversies. Scientific knowledge is always developing and changing. What is received as true today may well be overturned and shown to be false in the future. For this reason, open-minded assessment of evidence is a key virtue of scientific reasoning. Therefore, failing to teach both sides of the evidence in the climate change dispute is indoctrinating, even if the AGCC thesis turns out to be true, or so conclude the critics.

In presenting their evidence that there is no scientific consensus about the AGCC thesis, critics of the NGSS are claiming that any education presenting only this side of the argument would constitute a biased presentation of the evidence. This would be indoctrinating if it resulted in dogmatic adherence to the NGSS view.

The Indoctrination Charge Refuted

Science teaching that conforms to the NGSS weather and climate standards would not indoctrinate students to the AGCC thesis on even a weak standard. To be indoctrinating on a weak standard, teaching to these standards would have to meet one condition: we would need to have reason to expect that it will result in students who *dogmatically* believe that human activity is a significant cause of observed global climate change. Do we?

For at least a few reasons, we do not. Some students, perhaps many, are likely not to believe the AGCC thesis at all. Others might believe it dogmatically, but not for reasons especially attributable to the NGSS. For different reasons, still others would believe it. But they would not believe it dogmatically, even if their beliefs end up conforming entirely to the NGSS "party line."

Though in many states, examples of lessons satisfying the standards are given, the NGSS do not fully specify the content of educational curricula; schools and teachers have some latitude to determine what educational content suffices. I will assume in what follows that curricula conforming to the NGSS, expertly delivered, would enable students to understand the rationale behind the standards. That is, it would enable students to see, work with, and evaluate the scientific ideas, methods, and evidence that led the NGSS authors to include the standards they did (and perhaps to understand why they excluded certain ideas from the standards, for example, creationism as an alternative to the theory of evolution). This conception is consistent with the fundamental aims of the NGSS movement, namely, to reform the teaching of science to enable students to think like scientists.

Relative to the teaching of climate change, this conception requires that teachers go over the evidence relating to the occurrence of climate change and its natural and anthropogenic causes. Let us assume then that they do. My first claim—alas—is that many students receiving this education are likely to finish school not believing the AGCC thesis (and thus not having been indoctrinated to believe it). The main reason is that science literacy, potentially including literacy in specifically climate science, is not clearly the strongest determinant of a person's climate beliefs. The extent to which new information about climate change can influence a person's beliefs about the reality or causes of climate change is controversial. Some researchers find that individuals who learn to describe the basic mechanism of climate change in even a few sentences experience an increase in their confidence that climate change is real.[33] Educational attainment is also a strong predictor of belief in climate change.[34] If scientists hope to educate a scientifically literate public, we should hope that new information has at least some power to improve people's scientific beliefs.

However, it is widely believed (among science cognition researchers) that cultural and political identification strongly predict science beliefs, including beliefs about climate change. In a widely cited study, for example, Dan Kahan and colleagues found that self-reported knowledge of climate change is even *inversely* related to belief in the reality of climate change among conservatives.[35] That is, conservative subjects were more likely to deny that climate change is a problem the more they reported knowing about climate change.

In general, researchers increasingly find that information deficit is a poor explanation of climate change denialism.[36] Deniers reject the AGCC thesis not because they are ignorant of the science of climate change. They reject it because they feel that something central to their identity is threatened by accepting that science. If so, supplying them with more information (evidence) supporting the AGCC thesis will not change their beliefs.

Because political identification exerts such a powerful influence, it is likely that many students who receive an NGSS-conforming science education will still deny the AGCC thesis. They are likely to do so to the extent that they share any values and beliefs that lead their parents to deny climate change. For students sharing those values, the new scientific information they encounter in classrooms will be filtered through ideological skepticism and thus rejected.

By the same token, some students are liable to come away dogmatically believing the AGCC thesis. For these students, an NGSS-conforming education will confirm beliefs they held antecedently for reasons of political identity. However, the beliefs of these students, if held dogmatically, will not be dogmatic because they have received an NGSS-conforming education. On the contrary, that education would, if taken up by them, enable them to justify their belief with appropriate evidence and arguments. The beliefs of such students will be dogmatic because and to the extent that they fail to take up this evidence and argumentation, leaving them unable to justify their beliefs.

Thus, finally, some students will come away believing the AGCC thesis, but not dogmatically. These students will be those who digest the content of an NGSS-conforming science education and thus come to understand the fundamentals of climate science and the evidence supporting the AGCC thesis. These students will then be able to justify their beliefs in ways that are responsive to the evidence, that is, not dogmatically.

Some may still not be convinced. An NGSS-conforming climate education might end up being one-sided in either omitting denialist evidence against the AGCC thesis, or else being presented in such a way that students are led to see the errors in this evidence. How could this fail to produce dogmatic belief? Wouldn't avoiding dogmatism at least presuppose a neutral presentation of the evidence against one's beliefs and hearing both sides?

As a first consideration, recall that we are assuming a weak standard of indoctrination. We are assuming that teachers delivering NGSS-conforming curricula have no intention to indoctrinate any particular beliefs, nor will they use any manipulative teaching methods. They will faithfully present the best available evidence on climate change. And they will do so in such a way that students can inspect and evaluate that evidence for themselves, without being pressured to accept any of it by forces beyond that exerted by reason itself (that is, beyond the pressure exerted when they recognize that the evidence in fact logically supports certain conclusions and not others).

We may also assume that students receiving an NGSS-conforming education will not as a result become scientifically dogmatic in general. As discussed earlier, the NGSS are designed to improve the teaching of science for better student understanding of the methods of science and the nature of scientific knowledge. They will understand that curiosity, open-mindedness, and humility to evidence are central science virtues. And they will understand this because they will understand that scientific knowledge is changeable and probabilistic, collaborative and creative, and all the rest. They will also, however, understand that this does not mean that no science is ever settled—that is, established such that valuable new knowledge results from deploying it toward answering new questions, and not from perpetually reevaluating its fundamental basis.

As a second consideration, we should distinguish believing from believing dogmatically, and believing dogmatically from believing closed-mindedly. Obviously, to believe at all is not necessarily to believe dogmatically, which involves believing with unjustified certainty and with resistance to contrary evidence.[37] When a person believes something dogmatically, they believe it very strongly despite being unable to produce a correspondingly strong justification or to change their beliefs in response to evidence.

This is not quite the same as believing something closed-mindedly. That involves an unwillingness to possibly change one's mind in light of alternatives or new information.[38] That is a state of mind for which a person might well have a strong justification, both in everyday belief and in science. A well-educated person should remain closed-minded to alternatives when confronted with beliefs like $2 + 2 = 5$ or that the earth is flat. Closed-mindedness to alternatives to these beliefs can be supported by strong justifications, and schoolteachers should not teach the alternatives to these beliefs. This is the case despite the existence of a robust community of persons who, on the basis of evidence they have carefully evaluated for themselves, believe that the earth is flat.[39] It is not the case that whenever there are multiple sides to some science issue, it is indoctrinating to leave out any one of them.

Admittedly, the justifications of many of our beliefs might turn out to be fairly complex—even beyond our ability to fully produce—and not universally accepted. We can demonstrate basic arithmetic truths very informally (for example, by taking two rocks and another two rocks and seeing that they make four). But formal proofs in mathematics of even simple truths are quite technical; most people lack the sophistication to do or follow them. Similarly, in the case of the shape of the earth, we can appeal to informal observational evidence, such as the shape of the shadow of the earth on the moon during a lunar eclipse. But we may lack access to other data, such as the shifting location of constellations on the horizon from different locations on the earth, or lack the sophistication to evaluate geometrical arguments about the positions of stars.[40] The best we might be capable of is to point to the famous NASA photograph "Earthrise"—taken by the Apollo 8 crew from behind the horizon of the moon on December 24, 1968—and say "There, see. It's round."

This justification should suffice to convince most people that our belief is justified and not a mere piece of dogmatism. But it likely won't persuade skeptics like flat-Earthers. It will not persuade them because it relies (ultimately) on "testimonial evidence," and not on any evaluation of evidence involving astronomical, mathematical, or other reasoning. Testimonial evidence is evidence we have based on testimony, that is, on what a trusted source tells us. Despite enabling us to see the shape of the earth—and thus also being perceptual evidence—Earthrise is testimonial evidence. It is a copy of what the Apollo 8 astronauts saw and does not constitute our actually seeing the earth from a distance for ourselves.

In this regard this piece of evidence relies on the trust we place in NASA and the Apollo 8 astronauts, including trust that they would not lie to us, that they actually did go to the moon, that they actually took this photograph rather than made it up, and that it faithfully copies what they saw. This trust is precisely what flat-Earthers reject, perhaps in part based on some extravagant conspiracy theory according to which the lunar landings were faked. They may also present us with some complex astronomical argument that we are unable to refute.

This does not mean that our belief that the earth is round is dogmatic. On the contrary, many of our fundamental beliefs are justified on the basis of trusted testimony. Moreover, they could only be justified in this way, since we either lack access to the relevant information or lack the capacity/skill to evaluate it. This does not make them less justified. Our decisions about which sources of information to trust need not themselves be arbitrary and can be justified.

While the subject of credibility is complex, in general we evaluate the credibility of a source by reference to the reliability of the source with respect to the truth (can the source be trusted to typically form true beliefs on the subject at hand?) and by reference to the honesty of the source in communicating the truth to us (does this source have any reason to deceive us?).[41] Sources of scientific information like NASA, university scientists, and the IPCC are widely trusted with respect to truth because only highly credentialed scientists having extensive knowledge and expertise qualify for positions at such places. They are trusted to be honest in conveying information in part because they have a reputation to uphold and because they are public institutions, subject to public scrutiny and, indeed, control should they be found to have an agenda. Thus, when support for flat-Eartherism cannot be found at NASA or university departments of astronomy (or similar institutions), but only in obscure forums on the internet, we rightly remain closed-minded to flat-Earther evidence, and science teachers (and standards documents) rightly omit flat-Earther evidence in classrooms.

The NGSS for weather and climate align with received wisdom about climate change as established by the most credible sources of climate information, including the IPCC, as well as surveys of earth scientists and other scientists.[42] Thus, students receiving an NGSS-conforming climate education would not have been indoctrinated if, as a result of normal instruction in science adhering to the NGSS, they come to believe, even closed-mindedly, that human activity is a significant cause of climate change. Such closed-minded belief would not be unjustifiable, and therefore not dogmatic.

The Incredibly Discreditable NIPCC

NGSS critics will obviously reject this argument. In their view, an NGSS-conforming education in itself is biased so that the beliefs of any student subject to it will be held dogmatically. It is biased in their opinion because it is premised on evidence that places trust in the wrong hands, namely, the IPCC and the scientific consensus about climate change.

We should not credit these claims, and teachers should not teach them. Critics like Idso and colleagues lay out many arguments on behalf of their supposed side of the issue.[43] However, many repeat debunked scientific claims, such as that there has been no global warming since 1998; that greenhouse gases could not be a significant cause of climate change because historical emissions of greenhouse gases follow upon rather than precipitate global warming; or that the likely environmental effects of climate change and global warming will be more beneficial than harmful for humans and the planet.[44]

Veritably, these scientific issues are arguably outside the expertise of the authors of *Why Scientists Disagree*. At the time during which the text was written, the authors collectively held doctorates in fields peripheral to climatology, had published little in focal peer-review journals of climatology, and held no academic appointments while receiving monetary compensation for their work on behalf of the Heartland Institute.[45] Moreover, the partisan political nature of the document they produced is evident from the opening sentences of the foreword, written (in 2015) by Marita Noon, a columnist for the hyperpartisan Breitbart.com:

> President Barack Obama and his followers have repeatedly declared that climate change is "the greatest threat facing mankind." This, while ISIS is beheading innocent people, displacing millions from their homeland, and engaging in global acts of mass murder.
>
> If it weren't so scary, it would be laughable. These statements should ring alarm bells in the minds of all Americans. They show how out of touch this president and the movement he leads are with reality and the American public.[46]

This is a bizarrely political opening statement for a putatively scientific report.

Other arguments that Idso et al. advance attempt to debunk the social-science research finding that scientists—including climatologists—agree that human activity is the primary driver of observed recent climate change. For example, they reject the results of Oreskes for failing to focus on the opinion of scientists claiming expertise specifically in climatology.[47] Meanwhile they reject the results of Doran and Zimmerman for failing to include the opinions of a broader range of scientists having relevant expertise *not* specifically within climatology.[48] Apparently Idso et al. have no principled view of the experts whose opinions really matter. Rather the relevant experts vary by the rhetorical purposes needed in the particular context, so that those whose opinions are *not* considered by any particular study revealing a consensus are the relevant experts.

It is richly ironic that three paid shills for the fossil fuel sector should thus accuse the IPCC—an international body whose work is produced by hundreds of volunteer scientists—of being political.

Conclusion

A science education conforming to the NGSS for weather and climate would not indoctrinate students to the AGCC thesis, even on a weak standard. However, one registering the fictitious dissent to the climate consensus might. The evidence about what experts in climatology believe is not on the side of the denier. Resistance to the consensus is spitting into the evidential wind and dogmatic if any position on this matter is.

Nevertheless, and as mentioned in an earlier section, implementation of the NGSS has been uneven. According to the National Science Teaching Association, twenty states have adopted the NGSS.[49] Another twenty-four have developed their own standards based on the same framework documents behind the NGSS. These numbers mask deep hidden differences in adoption of the standards that explicitly identify human activity as a cause of climate change, including in states in the Midwest. Michigan is listed among the states having adopted the NGSS. Missouri is listed among those having independent standards based on the same framework. However, neither standard HS-ESS2.D3 nor HS-ESS2.D4 (see the earlier discussion) appear in the standards for either state. Instead we find relatively vague standards, such as "Ask questions to clarify evidence of the factors that have caused the rise in global temperatures over the past century" (Michigan's middle school ESS3–5) and "Predict how human activity affects the relationships between Earth systems in both positive and negative ways" (Missouri's high school standard ESS3.D.2).

This state of affairs is likely the result of ongoing politicization of the climate change issue. Teachers remain wary of student or parent backlash, going so far as to avoid the words "climate change" in teaching the subject.[50] Whatever the cause, educational standards cannot indoctrinate with ideas they do not contain.

Indoctrination can of course be accomplished via omission. Teaching that avoids certain issues or certain evidence about those issues can have the effect, even if unintended, of shaping students' receptivity to evidence, with the result that they become dogmatic. Indeed, science education researchers have questioned the ethical content and perspectives latent in the NGSS, especially as they pertain to sustainability. The NGSS devote a whole set of standards at the high school level to "Human Sustainability." And the concept of sustainability appears throughout many other components of the standards.

Sustainability has been a famously contentious source of political disagreement, among both advocates and critics of the very concept. Critics of the concept include some environmentalists who take issue with its anthropocentric focus, that is, how the concept prioritizes the long-term social or economic interests of humans. For many environmentalists, an ethical relationship to the natural environment is not possible when humans take themselves to be the only animals who count from a moral point of view. For these environmentalists, a sustainable human existence will not be achieved through strategic thinking in which humans figure out how to "get their own" without spoiling the nest of resources on which their existence depends. Rather, it will come from establishing a more respectful, caring, or integrated relationship to the environment and the other life within it, one in which the good of humans is achieved alongside their good. It will be achieved, in short, by developing a *nonanthropocentric* way of being for humans.[51]

Global justice thinkers also take issue with the concept of "the human" in sustainability, though in a different way. While nations of the developed world—such as the United States and the European Union—can fairly be blamed for many problems of global sustainability, the impacts of these problems are not distributed evenly across the globe. Some of the worst problems of climate change, such as droughts or flooding, are in fact expected to impact parts of the developing world or the Global South more severely.[52] Assimilating all problems of sustainability to problems of human sustainability tends to efface, or obscure, these inequalities. This is a significant potential injustice; the developing world cannot equally be blamed for *causing* problems of global sustainability.

Given the title of the NGSS section on sustainability ("Human Sustainability"), one might reasonably suspect that such concerns do not appear. In an analysis of the concept of sustainability in the NGSS, researchers Noah Feinstein and Kathryn Kirchgasler in fact find that the "vision of sustainability [in the NGSS] resembles *ecological modernization*, a technology-centered, managerial perspective on sustainability" in which "'industrial innovation encouraged by a market economy and facilitated by an enabling state will ensure environmental conservation.'"[53] The NGSS evince "a faith in the power of technical progress" to ensure human survival, never mind the role such "progress" has itself played in giving rise to problems of global sustainability, or the impact of this progress for other humans living elsewhere or other (nonhuman) life on earth.[54] Such faith is reflected, for example, in standard HS-ESS3.D1: "Though the magnitudes of human impacts are greater than they have ever been, so too are human abilities to model, predict, and manage current and future impacts." In other words, don't worry your pretty little head about all the damage we are doing to the planet; we can just fix things once they are strained to the point of breaking.

However explicitly or implicitly these "techno-optimist" values appear in the NGSS, their presence is notable given their conservatism (as a form of environmental ethics) alongside the total absence of their alternatives. In the broader public of modern-day America, non-anthropocentric environmentalist or global justice ideals are not easily encountered. By contrast, the ideal of better living through scientific/technological progress is easily encountered. At the time of writing, it is known that certain simple behavior changes—mask wearing and social distancing—are effective in limiting the spread of COVID-19. However, the focus of the political administration (Donald Trump's) is not these changes—which indeed they have openly flouted, even mocked—but the development of vaccines and treatments. Techno-optimism is everywhere in modern American life. The culture of the Midwest is no exception, cradle that it is to the auto industry, mass monocultural farming, and quintessential American cities like Chicago. If children are to encounter any diversity of ethical thinking about such matters, schools will have to introduce it.

Thus, the presence of status quo environmental attitudes in the NGSS would represent a significant educational failing of those standards. Unlike the dispute about the AGCC thesis, there is more than one credible view about global justice and the right human relationship to the environment. If science teaching conforming to the NGSS makes credible alternatives harder to see—or worse, reinforces the side of the status quo—it is liable to produce students who adhere dogmatically to status quo ethical beliefs with respect to the environment and global justice. If so, the NGSS may well be indoctrinating, just not in the way critics have claimed.

Acknowledgments

I wish to thank Ian Smith, Lauren Bialystok, and Charles Hayes for helpful feedback on earlier drafts of this chapter.

NOTES

1. John Cook et al., "Consensus on Consensus: A Synthesis of Consensus Estimates on Human-Caused Global Warming," *Environmental Research Letters* 11, no. 4 (April 2016): 048002; William R. L. Anderegg et al., "Expert Credibility in Climate Change," *Proceedings of the National Academy of Sciences* 107, no. 27 (July 6, 2010): 12107–12109; Naomi Oreskes, "Beyond the Ivory Tower: The Scientific Consensus on Climate Change," *Science* 206, no. 5702 (December 2004): 1686; Peter T. Doran and Maggie Kendell Zimmerman, "Examining the Scientific Consensus on Climate Change," *Eos* 90, no. 3 (January 20, 2009): 22–23.

2. Katie Worth, "Mailings to Teachers Highlight a Political Fight over Climate Change in the Classroom," *Frontline/PBS*, March 23, 2018; Michael Melia, "School Lessons Targeted by Climate Change Doubters," *Spokesman-Review*, March 6, 2019; National Research Council, *Next Generation Science Standards: For States, By States*, 2013, https://doi.org/10.17226/18290.

3. Melia, "School Lessons"; Craig D. Idso, Robert M. Carter, and Fred D. Singer, *Why Scientists Disagree: The NIPCC Report on Scientific Consensus*, 2nd ed. (Arlington Heights, IL: Heartland Institute, 2016); David R. Legates et al., "Climate Consensus and 'Misinformation': A Rejoinder to Agnotology, Scientific Consensus, and the Teaching and Learning of Climate Change," *Science & Education* 24, no. 3 (April 1, 2015): 299–318.

4. Cook et al., "Consensus on Consensus."

5. Norman Lederman, Allison Antink-Meyer, and Stephen Bartos, "Nature of Science, Scientific Inquiry, and Socio-Scientific Issues Arising from Genetics: A Pathway to Developing a Scientifically Literate Citizenry," *Science & Education* 23, no. 2 (2014): 285–302.

6. "History of the IPCC," IPCC, https://www.ipcc.ch/about/history/.

7. Robert Pennock, *An Instinct for Truth* (Cambridge, MA: MIT Press, 2019).

8. The NGSS attempt to achieve this goal by organizing K–12 grade scientific information around a delimited set of "disciplinary core ideas," such as those in thermodynamics; "scientific and engineering practices," such as modeling; and "cross-cutting concepts," such as cause and effect, which are consistent across the grade levels.

9. National Research Council, *Next Generation Science Standards*.

10. Doran and Kendell Zimmerman, "Examining the Scientific Consensus."

11. In surveys of American public opinion, the Midwest is a somewhat amorphous region. Americans have less clear ideas about which states make it up than about which states make up the South. About a quarter of respondents to one survey place Oklahoma and Kentucky in the Midwest; another 10 percent go so far as to include Wyoming on the western border and West Virginia on the eastern. For discussion, see Walt Hickey, "Which States Are in the Midwest?" *FiveThirtyEight*, April 29, 2014, https://fivethirtyeight.com/features/what-states-are-in-the-midwest/; Walt Hickey, "Which States Are in the South?" *FiveThirtyEight*, April 30, 2014, https://fivethirtyeight.com/features/which-states-are-in-the-south/.

12. All claims and figures to follow are supported by research conducted by the Yale Program on Climate Change Opinion and are gleaned from Jennifer Marlon et al., "Yale Climate Opinion Maps 2020," Yale Program on Climate Change Communication, September 2, 2020, http://climatecommunication.yale.edu/visualizations-data/ycom-us/.

13. Leo Hickman, "Leaked Heartland Institute Documents Pull Back Curtain on Climate Scepticism," *The Guardian*, February 15, 2012, sec. Environment, http://www.theguardian.com/environment/blog/2012/feb/15/leaked-heartland-institute-documents-climate-scepticism.

14. Idso, Carter, and Singer, *Why Scientists Disagree*.

15. Hickman, "Leaked Heartland Institute Documents."

16. The Heartland Institute, https://www.heartland.org.

17. Eric Plutzer et al., "Climate Confusion among U.S. Teachers," *Science* 351, no. 6274 (February 12, 2016): 664–665; Caroline Halter, "Requiring Schools to Teach Climate

Change Risks Backlash in Oklahoma," *StateImpact Oklahoma*, July 11, 2019, https://stateimpact.npr.org/oklahoma/2019/07/11/requiring-schools-to-teach-climate-change-risks-backlash-in-oklahoma/.

18. IPCC, *Climate Change 2014: Synthesis Report. Contribution of Working Groups I, II, and III to the Fifth Assessment Report of the Intergovernmental Panel on Climate Change*, 2015, https://www.ipcc.ch/report/ar5/syr/.

19. Christopher Thomas Holland, "The Implementation of the Next Generation Science Standards and the Tumultuous Fight to Implement Climate Change Awareness in Science Curricula," *Brock Education Journal* 29, no. 1 (February 3, 2020): 35–52.

20. Randall Curren and Ellen Metzger, *Living Well Now and in the Future: Why Sustainability Matters* (Cambridge, MA: MIT Press, 2017).

21. Charles Fletcher, *Climate Change: What the Science Tells Us*, 2nd ed. (Hoboken, NJ: Wiley, 2019).

22. All data again are culled from Marlon et al., "Yale Climate Opinion Maps 2020."

23. Fletcher, *Climate Change*, 267.

24. "Kansas Drought Information," United States Department of Agriculture, Natural Resources Conservation Service Kansas, https://www.nrcs.usda.gov/wps/portal/nrcs/detailfull/ks/water/?cid=nrcs142p2_033642.

25. "Governor Declares Drought Emergency, Warnings, and Watches for Kansas Counties," Kansas: Office of the Governor, July 6, 2020, https://governor.kansas.gov/governor-declares-drought-emergency-warnings-and-watches-for-kansas-counties/.

26. Fletcher, *Climate Change*, 268.

27. Ambarish Vaidyanathan et al., "Heat-Related Deaths—United States, 2004–2018," *CDC Morbidity and Mortality Weekly Report* 69, no. 24 (June 19, 2020): 729–734.

28. J. P. White, "Indoctrination," in *The Concept of Education*, ed. R. S. Peters (London: Routledge & Kegan Paul, 1967).

29. Eamonn Callan and Dylan Arena, "Indoctrination," in *The Oxford Handbook of Philosophy of Education*, ed. Harvey Siegel (Oxford: Oxford University Press, 2009).

30. All of these are discussed in I. A. Snook, *Indoctrination and Education*, new ed. (London: Routledge & Kegan Paul, 1975).

31. For a similar example and further discussion of examples of this sort, see Rodger Beehler, "The Schools and Indoctrination," *Journal of Philosophy of Education* 19 (1985): 261–272; and Callan and Arena, "Indoctrination."

32. Idso, Carter, and Singer, *Why Scientists Disagree*, xii.

33. Michael Andrew Ranney and Dav Clark, "Climate Change Conceptual Change: Scientific Information Can Transform Attitudes," *Topics in Cognitive Science* 8, no. 1

(January 1, 2016): 49–75.

34. David J. Hess and Alexander Maki, "Climate Change Belief, Sustainability Education, and Political Values: Assessing the Need for Higher-Education Curriculum Reform," *Journal of Cleaner Production* 228 (August 10, 2019): 1157–1166.

35. Dan M. Kahan et al., "The Polarizing Impact of Science Literacy and Numeracy on Perceived Climate Change Risks," *Nature Climate Change* 2, no. 10 (October 2012): 732–735.

36. Adrian Bardon, *The Truth about Denial* (Oxford: Oxford University Press, 2019), chap. 2.

37. Bob Altemeyer, "Dogmatic Behavior among Students: Testing a New Measure of Dogmatism," *Journal of Social Psychology* 142, no. 6 (December 2002): 713–721.

38. Heather Battaly, "Can Closed-Mindedness Be an Intellectual Virtue?" *Royal Institute of Philosophy Supplement* 84 (2018): 23–45.

39. Rob Picheta, "The Flat-Earth Conspiracy Is Spreading around the Globe. Does It Hide a Darker Core?" *CNN*, November 18, 2019, https://www.cnn.com/2019/11/16/us/flat-earth-conference-conspiracy-theories-scli-intl/index.html.

40. Such as Aristotle discusses in *De Cealo* (*On the Heavens*), trans. J. L. Stocks, bk. II:14, http://classics.mit.edu/Aristotle/heavens.mb.txt. Thanks to Ian Smith for pointing this out.

41. Miranda Fricker, "Rational Authority and Social Power: Towards a Truly Social Epistemology," *Proceedings of the Aristotelian Society* 98 (1998): 159–177.

42. Cook et al., "Consensus on Consensus"; Anderegg et al., "Expert Credibility in Climate Change"; Doran and Kendell Zimmerman, "Examining the Scientific Consensus."

43. Idso, Carter, and Singer, *Why Scientists Disagree*.

44. Detailed discussion of all these claims can be found in Fletcher, *Climate Change*.

45. Hickman, "Leaked Heartland Institute Documents."

46. Idso, Carter, and Singer, *Why Scientists Disagree*, xi.

47. Oreskes, "Beyond the Ivory Tower"; Idso, Carter, and Singer, *Why Scientists Disagree*, 12.

48. Doran and Kendell Zimmerman, "Examining the Scientific Consensus"; Idso, Carter, and Singer, *Why Scientists Disagree*, 14.

49. "About the Next Generation Science Standards," National Science Teaching Association, https://ngss.nsta.org/About.aspx.

50. Ines Kagubare, "Some States Still Lag in Teaching Climate Science," *Scientific American*, February 8, 2019.

51. For a sophisticated version of this kind of view, see Paul W. Taylor, *Respect for Nature: A Theory of Environmental Ethics* (Princeton, NJ: Princeton University Press, 1986).

52. Fletcher, *Climate Change*.

53. Noah Weeth Feinstein and Kathyrn Kirchgasler, "Sustainability in Science Education? How the Next Generation Science Standards Approach Sustainability, and Why It Matters," *Science Education* 99, no. 1 (2015): 121–144, at 134, emphasis original, citing A. Blowers, "Environmental Policy: Ecological Modernisation of the Risk Society?" *Urban Studies* 34, no. 5–6 (1997): 845–871, 847.

54. Feinstein and Kirchgasler, "Sustainability in Science Education?," 134.

SUGGESTED READINGS

Feinstein, N. W., and D. Waddington. "Individual Truth Judgments or Purposeful, Collective Sensemaking? Rethinking Science Education's Response to the Post-truth Era." *Educational Psychologist* 55, no. 3 (2020): 155–166.

Ferkany, Matt, and K. P. Whyte. "The Compatibility of Liberalism and Mandatory Environmental Education." *Theory and Research in Education* 11, no. 5 (2013): 5–21.

Taylor, Rebecca. "Indoctrination and Social Context: A System-Based Approach to Identifying the Threat of Indoctrination and the Responsibilities of Educators." *Journal of Philosophy of Education* 51, no. 1 (2017): 38–58.

About the Contributors

J. M. Dieterle is a professor of philosophy in the Department of History and Philosophy at Eastern Michigan University. She is the editor of *Just Food: Philosophy, Justice and Food* (2015) and has published articles on environmental philosophy in *Environmental Ethics*, the *Journal of Agricultural and Environmental Ethics*, and *Agriculture and Human Values*.

Justin Donhauser is an assistant professor in the Department of Philosophy at Bowling Green State University. He is an alum of the Rotman Institute of Philosophy and of the National Science Foundation's Ecosystem Restoration through Interdisciplinary Exchange program. Donhauser specializes in applied philosophy of science and environmental ethics. His primary research clarifies how research and methods in environmental sciences can inform public policy and resource management decision-making.

Matt Ferkany is an associate professor of philosophy and affiliated faculty of the Environmental Science and Policy Program at Michigan State University. Ferkany's work focuses on well-being and virtue, moral education, and environmental ethics and education. His work has been published in numerous peer-reviewed journals and supported by grants from the Spencer Foundation.

Joel MacClellan is an assistant professor of philosophy at Loyola University New Orleans. His area of expertise is ethics, especially environmental and animal ethics. His publications include articles in *Between the Species*, *Ethics & the Environment*, and the *Journal of Animal Ethics*. He has presented his research in the United States, Canada, Mexico, and the Netherlands. He is a returned Peace Corps Volunteer (Environment Program, Panama, 2003–2005). Dr. MacClellan was a scholar-in-residence at Wesleyan University as the New York University's Animal Ethics and Public Policy Fellow.

Matthew Meyer is an associate professor of philosophy at the University of Wisconsin–Eau Claire. He has long been interested in cross-historical and interdisciplinary studies in philosophy. He has presented papers on environmental philosophy, with a focus on ecophenomenology and its insight into the related problems of domination and consumption. He has also written multiple articles on Nietzsche. His interests are in phenomenology, existentialism, ecophenomenology, ecofeminism, environmental philosophy, and psychoanalysis.

Heather Ann Moody is the director and an associate professor of American Indian studies at the University of Wisconsin–Eau Claire. She is an enrolled member of the Ho-Chunk Nation. Her work centers around her background in teaching and learning in relation to the incorporation of accurate American Indian curriculum in K–12 and beyond, known in Wisconsin as Act 31. In addition to her campus work, she is a teacher exemplar and works with surrounding school districts to implement Act 31. Moody also serves as the acting director for the Center for Racial and Restorative Justice and is the co-advisor for the Inter-Tribal Student Council.

Benjamin J. Pauli is an associate professor of social science in the Department of Liberal Studies at Kettering University. He is the author *of Flint Fights Back: Environmental Justice and Democracy in the Flint Water Crisis* (2019), president of the Board of the Environmental Transformation Movement of Flint, vice-chair of the Flint Water System Advisory Council, and a member of the Flint Area Health and Environment Partnership. He also serves as a representative of the academic community on the U.S. EPA's National Environmental Justice Advisory Council. His interests include issues of water governance, environmental justice, and principles of engagement and collaboration in community-based research.

Ian A. Smith is an associate professor of philosophy, along with being the chair of the Philosophy and Religious Studies Department at Washburn University (PhD, University of Utah). Smith primarily teaches ethics and logic classes. Smith has published work in environmental ethics and normative ethics in *Environmental Ethics*, *Journal of Social Philosophy*, *Journal of Business Ethics*, and *Philosophia*, among other journals. Smith is the author of *The Intrinsic Value of Endangered Species* (2016).

William O. Stephens is a professor emeritus of philosophy at Creighton University. His books are *Marcus Aurelius: A Guide for the Perplexed* (2012), *Stoic Ethics: Epictetus and Happiness as Freedom* (2007), *The Person: Readings in Human Nature* (2006), and *The Ethics of the Stoic Epictetus: An English Translation, Revised Edition* of A. F. Bonhöffer's classic (2021). His work on Stoicism includes essays on refugees, love, death, animals, sportsmanship, travel, habit, ecology, *Star Wars*, and *Gladiator* (2000). He has also published on friendship, ethics and animals, and philosophical vegetarianism.

Levi Tenen is an assistant professor of philosophy in the Department of Liberal Studies at Kettering University. His work lies at the intersection of value theory, environmental philosophy, and aesthetics. He has published papers in *Environmental Ethics*, the *Journal of Value Inquiry*, and *Environmental Values*, and has served as coeditor for a special issue in the *Journal of Aesthetics and Art Criticism* on environmental aesthetics and ethics.

Wade Tornquist earned his PhD in physical chemistry from the University of Minnesota in 1986. He is a professor of chemistry at Eastern Michigan University, where he has also served for more than two decades as an academic administrator. His most recent role is administering graduate school and research development and support functions. He has publications in chemistry, physics, and surface science journals and monographs.